Springer Laboratory

Springer

Berlin
Heidelberg
New York
Barcelona
Hong Kong
London
Milan
Paris
Singapur
Tokyo

Sadao Mori · Howard G. Barth

Size Exclusion Chromatography

With 63 Figures and 21 Tables

 Springer

Prof. Sadao Mori
Mie University
4–8, Ibuka
Nagoya 453-0012
Japan

Dr. Howard G. Barth
DuPont Company
Central Research and Development
P.O. Box 80228
19880-0228 Wilmington, Delaware
USA

ISBN 978-3-642-08493-5

Libary of Congress Cataloging-in-Publication Data applied for

Die Deutsche Bibliothek – CIP-Einheitsaufnahme
Mori, Sadao:
Size exclusion chromatographie : with 20 tables / Sadao Mori ;
Howard G. Barth. – Berlin ; Heidelberg ; New York ; Barcelona ;
Hong Kong ; London ; Milan ; Paris ; Singapur ; Tokio :
Springer, 1999
  (Springer laboratory)

© Springer-Verlag Berlin Heidelberg 2010
Printed in Germany

The use of general descriptive names, registered names, etc. in this publication does not imply, even in the absence of a specific statement, that such names are exempt from the relevant protective laws and regulations and free for general use.

Product liability: The publisher cannot guarantee the accuracy of any information about dosage and application contained in this book. In every individual case the user must check such information by consulting the relevant literature.

Coverdesign: Erich Kirchner, Heidelberg

Preface

The molecular weight (MW) and molecular weight distribution (MWD) of polymers are perhaps their most important characteristics, governing both physical properties and end-use applications. As a result, MW and MWD are needed to establish structure-property-processing relationships for the development of new products and the improvement of existing materials. Furthermore, these measurements are needed to access the quality of a material, and for process control and monitoring polymerization kinetics.

Size exclusion chromatography (SEC), also referred to as gel permeation chromatography or gel filtration, is the most widely used method for determining MW and MWD. Almost every manufacturer or laboratory involved with polymeric materials is or should be involved with this analytical technique. Unfortunately, the theory and practice of SEC is not a customary part of one's analytical training, and must be learned on the job. Furthermore, during the past several years, there have been significant advances in SEC column technology, detection systems, and new applications. In fact, SEC can now be used, in conjunction with molecular-weight-sensitive detection systems, for determining absolute MW/MWD, as well as molecular structure, such as molecular size, conformation, and branching, as a function of MW. Furthermore, by interfacing SEC with spectrometry, polymer compositional heterogeneity can be determined.

This book is written for chemists and technicians who have little or no background in SEC, and who would like to gain a solid understanding of the fundamentals and practice of SEC. For those who are more experienced in SEC, this volume should serve as a review and resource of SEC basics and advanced developments. This volume introduces the reader to the fundamentals of SEC with emphasis on the practical aspects of the technique, such as column and mobile phase selection, calibration, detector capabilities, and offers guidelines for performing SEC on many types of polymers, especially those of industrial importance.

The authors wish to thank the Department of Industrial Chemistry, Mie University and Central Research and Development of the DuPont Company for supporting this project.

Nagoya, Japan   Sadao Mori
Wilmington, Delaware, USA   Howard G. Barth
March 1999

Table of Contents

Abbreviations

AD	analog-to-digital
BHT	2,6-di-*tert*-butyl-*p*-cresol
CCD	chemical composition distribution
CV	capillary viscometer
DALLS	dual-angle laser light scattering
DMF	*N*,*N*-dimethylformamide
DMSO	dimethylsulfoxide
DP	degree of polymerization (also **d**)
EDTA	ethylenediaminetetraacetate
EP	ethylene-propylene
EVA	ethylene-vinyl acetate
FTIR	Fourier-transform infrared
GFC	gel filtration chromatography
GPC	gel permeation chromatography
HA	hyaluronic acid
HDPE	high-density polyethylene
HETP	height equivalent to a theoretical plate
HFIP	hexafluoroisopropanol
HPLC	high-performance liquid chromatography
HPSEC	high-performance size exclusion chromatography
HTSEC	high-temperature size exclusion chromatography
IR	infrared
iPP	isotactic polypropylene
LALLS	low-angle laser light scattering
LCB	long-chain branching
LDPE	low-density polyethylene
LLDPE	linear low-density polyethylene
LS	light scattering
MALLS	multi-angle laser light scattering
MEK	methyl ethyl ketone
MW	molecular weight
MWD	molecular weight distribution
M_n	number-average molecular weight
M_w	weight-average molecular weight
M_z	z-average molecular weight
N	number of theoretical plates in a column
NaPSS	sodium poly(styrene sulfonate)
NMP	*N*-methylpyrrolidone

OCP	*o*-chlorophenol
ODCB	*o*-dichlorobenzene
OEG	oligoethylene glycol
PAA	polyamic acid
PAN	poly(acrylonitrile)
PEEK	poly(ether ether ketone)
PEG	polyethylene glycol
PEO	poly(ethylene oxide)
PET	poly(ethylene terephthalate)
PFP	pentafluorophenol
PMMA	poly(methyl methacrylate)
PPS	poly(phenylene sulfide)
PS	polystyrene
PVA	poly(vinyl alcohol)
PVB	poly(vinyl butyral)
PVC	poly(vinyl chloride)
P2VP	poly(2-vinylpyridine)
RALLS	right-angle laser light scattering
RI	refractive index
SCB	short-chain branching
SDS	sodium dodecyl sulfate
SEC	size exclusion chromatography
TALLS	triple-angle laser light scattering
TCB	1,2,4-trichlorobenzene
TFA	trifluoroacetic acid
TFE	2,2,2-trifluoroethanol
THF	tetrahydrofuran
UV	ultraviolet region
VIS	visible region
VPO	vapor pressure osmometry

1 Introduction

1.1 Molecular Weight Measurements

For polymeric materials, the molecular weight (MW) or molecular *size* plays a critical role in determining the mechanical, bulk, and solution properties of polymeric materials, as shown in Table 1.1. These properties govern polymer processing, end-use performance, and interactions with neighboring molecules. As compared to small molecules, which have a discrete and well-defined MW, most synthetic polymers are composed of hundreds to thousands of chains of different MWs that result in characteristic molecular weight distributions (Fig. 1.1). Each polymer will have, in essence, a distinctive molecular weight distribution (MWD), and its shape and breadth will depend on the polymerization mechanism, kinetics and conditions. In addition to synthetic polymers, many naturally occurring polymers, such as petroleum-based products, lignins, natural oils and fats, humic acids, natural rubber, cellulosics, and polysaccharides have characteristic MWDs, depending on their source and method of isolation. Nucleic acids and most proteins, however, consist of a well-defined molecular structure of a single-molecular-weight species.

In order to define the MWD of a polymer, moments or statistical averages of the distribution are calculated, as discussed in greater detail in Chap. 6. As shown in Fig. 1.2, typically three moments are calculated: number-average M_n, weight-average M_w, and z-average M_z molecular weights. The magnitude of M_n is sensitive to the presence of low-molecular-weight material, M_w is sensitive to high-molecular-weight components, while M_z reflects changes in the very high-molecular-weight portion of the distribution. The width of the distribution, called the polydispersity, is determined from the ratio of M_w/M_n or, in some ca-

Table 1.1. Polymer properties that are influenced by MW or MWD

Processability	Glass-transition temperature
Solution viscosity	Hardness
Melt viscosity	Tear strength
Tensile strength	Stress-crack resistance
Brittleness	Impact resistance
Flex life	Stress relaxation
Toughness	Creep strain
Drawability	Compression
Tackiness	Fatigue
Gas permeability	Wear

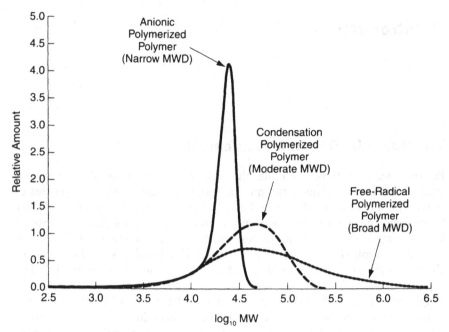

Fig. 1.1. Representative MWDs of three common types of synthetic polymers. The MWDs are from actual samples

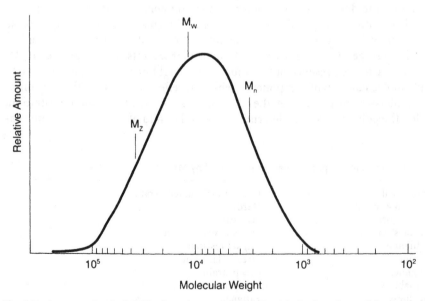

Fig. 1.2. An example of a MWD of a polymer showing the relative locations of the number- (M_n), weight- (M_w) and z-average (M_z) MWs or statistical moments

ses, M_z/M_w. If the polymer consists of a single-molecular-weight component, it is considered to be monodisperse.

Values of M_n, M_w, and M_z can be measured independently by physical methods [1–4]. M_n can be determined by end-group analysis or by colligative properties, such as boiling point elevation, freezing point depression, membrane osmometry, or vapor pressure osmometry, although the two latter methods are preferred. End-group analysis is useful mainly for condensation polymers or polymers with defined chain ends. For example, polyamides that have a stoichiometric amount of carboxyl and amino chain ends can be measured by titration, the upper M_n limit of which is approximately 10^4 g/mol.

The relationship between colligative properties and MW for infinitely dilute solutions is based on the fact that the solute activity is equal to its mole fraction. In the case of membrane osmometry, the osmotic pressure π that is developed across a semipermeable membrane separating a polymer solution from the pure solvent is related to M_n by

$$\pi = RT\left((c/M_n) + A_2 c + A_3 c^2 + \ldots\right) \tag{1.1}$$

where c is the solute concentration, A_2 and A_3 are the 'osmotic' second and third virial coefficients, T is temperature, and R is the gas constant. At infinite dilution, the virial coefficients approach zero, and M_n is calculated from

$$\pi/c = RT/M_n \tag{1.2}$$

Because of the semipermeable nature of membranes, only polymer molecules greater than the molecular weight cut-off of the membrane will contribute to the osmotic pressure; polymers that are smaller than the cut-off will diffuse across the membrane and will not influence π. Typically, the MW range that can be determined with membrane osmometry is $10^4 - 5 \times 10^5$ g/mol.

When the vapor phase of a polymer solution is in equilibrium with the saturated vapor phase of the solvent, the vapor pressure of the solvent component of the solution P will be less than the vapor pressure of the pure solvent P°, and their ratio can be expressed as

$$\ln(P/P^\circ) = V^\circ\left((c/M_n) + A_2' c + A_3' c^2 + \ldots\right) \tag{1.3}$$

where V° is the molar volume of the solvent, and A_2' and A_3' are the 'vapor pressure' second and third virial coefficients. Because of the experimental difficulties involved in measuring the vapor pressure difference between the solution and solvent, a different approach is used, called vapor pressure osmometry (VPO). Rather than determining the vapor pressure difference, the temperature difference which results from the different rates of solvent evaporation from a droplet of pure solvent (T°) and condensation of the solvent onto a droplet of a polymer solution (T) is measured. In practice, both solvent and polymer solution droplets are supported on respective thermistors in an atmosphere of pure solvent. The temperature difference at steady state is

$$(T - T^\circ)/c = K_s/M_n \tag{1.4}$$

where K_s is a calibration constant. In general, the M_n upper limit of VPO is 10^4 g/mol.

M_w can be determined from light scattering, in which Rayleigh light scattering from a polymer solution is measured as a function of the scattering angle and the polymer concentration. By extrapolation of the excess scattered light intensity to zero angle and infinite dilution, M_w can be determined. (The theory and application of light scattering detectors are discussed in detail in Chap. 8). The typical MW range covered by light scattering is $> 10^3$ g/mol. Although seldom used nowadays, ultracentrifugation can also be employed to measure M_w, M_z and MWD; however, because of problems regarding detection sensitivity, length of analysis, data interpretation, expense, and instrumentation maintenance, it is used mainly as a research tool for investigating the hydrodynamic properties of proteins.

One of the major limitations in determining average MW values is that the overall *shape* of the MWD remains unknown. Thus, two polymers may have the same average MW values but completely different MWDs. Furthermore, knowledge of an average MW gives absolutely no information regarding the modality of the distribution or the presence of shoulders or other characteristic features which may ultimately affect polymer performance. In view of this drawback, a technique is needed that can not only measure average MW values but can also give the MWD.

1.2 Brief Description of Size Exclusion Chromatography

By far the most popular and convenient method of determining the average molecular weight and the MWD of a polymer is size exclusion chromatography (SEC). Typically, in less than 30 min using standard high-performance liquid chromatography (HPLC), the complete MWD of a polymer can be determined along with *all* of the statistical moments of the distribution. Thus, SEC has essentially supplanted many classical MW techniques for routine measurements. (The term 'gel permeation chromatography' (GPC) is also used, but refers to the separation of organosoluble polymers; whereas, the name 'gel filtration chromatography' (GFC) refers specifically to the separation of biopolymers in an aqueous mobile phase. Other terms, such as steric exclusion chromatography, permeation chromatography, or exclusion chromatography have been used. In this book, size exclusion chromatography is adapted which properly describes the separation mechanism [5].)

SEC separates on the basis of molecular hydrodynamic volume or size, rather than by enthalpic interactions with the stationary phase or packing surfaces, as is the case with other modes of liquid chromatography, such as adsorption, partition, or ion exchange. In SEC, the polymer is dissolved in an appropriate solvent and is injected into a column packed with porous particles of fairly defined pore size. The SEC mobile phase is generally the same solvent used to dissolve the polymer. As the polymer elutes through the column, molecules that are too large to penetrate the pores of the packing elute in the interstitial or void volume of the column. As the molecular size of the polymer decreases with respect to the average pore size of the packing, polymer molecules penetrate into the pores and access greater pore volume and, as a result, elute at a later time. Small molecules, which can freely diffuse into and out of the pores and can sample the total pore volume, elute at the total elution volume of the column. High-molecular-weight material elutes first from an SEC column, followed by low-molecular-weight components. Since SEC is a *relative* and not an *absolute* molecular weight technique, columns must be calibrated with polymer standards of known molecular weight (Chap. 7), or an online light scattering detector must be used (Chap. 8).

When compared to other modes of HPLC, the instrumentation requirements for SEC are somewhat simpler since mobile phase gradients are not used; however, computer support for data acquisition and processing is critical. Depending on the application, high-temperature capability, for example, 150 °C, may be required (see Chap. 10). Although any concentration-sensitive detector can be used in SEC, the differential refractometer is the most widely utilized (Chap. 3). Column calibration is of utmost concern in SEC, and a number of different approaches have been developed (Chap. 7), including the use of molecular-weight sensitive detectors (Chap. 8). As discussed in Chaps. 4 and 5, column and mobile phase selection are obviously essential to the success of an SEC separation.

1.3 Brief Historical Perspective

In 1959, Porath and Flodin [6] developed a cross-linked dextran gel which, when swollen in an aqueous media, was able to separate proteins on the basis of molecular weight. This gel was made available commercially under the name of Sephadex, and has been used extensively for biopolymer separation in low-pressure GFC systems using aqueous mobile phases. Five years later, Moore disclosed a procedure for measuring the MW and MWD of synthetic polymers soluble in organic media using porous, cross-linked polystyrene (PS) gels (copolymer of styrene and divinyl benzene), and coined the term 'gel permeation chromatography' [7].

With the column technology and equipment available at that time, SEC required many hours and sometimes days to achieve a separation. Initially particle diameters of SEC packings were larger than 30 µm with rather broad particle size distributions. Typically, a 4-ft column packed with PS gels generated approximately 400–700 theoretical plates per foot. Usually four columns of different pore sizes were connected giving about 6000–11 000 total plates. In the 1970s, rigid PS gel packings of 10 µm were introduced that had much narrower particle size distributions. Columns packed with this type of PS gel gave about 10 000 plates per foot, a 15-fold improvement in column efficiency. As a result, the total column length and analysis time were reduced by one-fifteenth using typical flow rates. In the 1990s, column packing technology was improved further with the introduction of packings having diameters of 3 and 5 µm.

Columns packed with these gels have 25 000 and 15 000 plates per foot, respectively. In practice, modern SEC is performed, in general, with two 1-ft columns of 30 000 to 50 000 theoretical plates with a separation time of less than 30 min.

1.4 Applications

The typical separation range of SEC is approximately 10^2 to 10^6 g/mol, thus small molecules, as well as high-molecular-weight solutes, can be size separated depending on the pore size of the packing. Although polymers of greater than 10^6 g/mol can be analyzed, care must be taken to avoid the onset of non-size exclusion effects, such as polymer shear degradation and concentration effects (Chap. 5). In principle, any type of sample can be analyzed by SEC provided that it can be solubilized and that there are no enthalpic interactions present between the sample and the packing material, as discussed in Chap. 2.

Complex, unknown mixtures can be screened by SEC to determine the presence of small molecules, oligomers and polymeric components. Although SEC peak capacity (the number of peaks that can be separated with near baseline resolution) is about an order-of-magnitude lower than in HPLC [8], SEC can be utilized in place of HPLC for the analysis of small molecules if there is sufficient molecular-weight differences among solutes. Under ideal SEC conditions, the entire sample should elute within one column volume, thus sample recovery is usually not a concern. Furthermore, because elution order is related to molecular size, peak identification is greatly facilitated. Using molecular weight calibrants, the molecular weight of eluting components can be approximated. Thus, SEC can be considered to be a chromatographic 'mass spectrometer', albeit of low resolution.

Preparative scale SEC is accomplished with large-diameter columns, in which mg-g quantities can be fractionated (Chap. 12). Moreover, sample cleanup can be achieved quite easily with SEC, in which low-MW components, such as pesticides, herbicides, or pollutants, can be isolated from complex matrices, as in the case of biological or environmental samples. Desalting biological compounds is typically done with SEC. For example, a protein solution can be separated from dissolved electrolytes with high recovery using water as the mobile phase.

Besides being used for measuring MWDs, SEC can be used simply to compare relative molecular-weight differences among samples *without* column calibration. In fact, many analytical laboratories use SEC for this type of problem solving when analyzing competitive products, or for comparing or ascertaining sample composition.

Physicochemical properties of biopolymers and synthetic polymers can be studied with SEC, such as association and ligand binding, polymerization kinetics, and protein folding and conformation studies. For synthetic polymers, long-chain branching and polymer conformation, for example, can be determined using MW-sensitive detectors (Chap. 8). One of the emerging application areas of SEC is the use of on-line spectroscopic detectors, such as Fourier transform infrared (FTIR), NMR and MS, for determining the compositional heterogeneity of polymeric materials (see Chaps. 3 and 12) [9].

1.5 General References

There have been a number of books written about SEC [10-12], including contributed volumes [13-17] and symposium proceedings [18-26]. Waters Corp. [27] also publishes proceedings of their International GPC Symposia, which are held biennially. Of interest also are books related to polymer fractionation and chromatography [28-34]. Standard SEC methods have been published by the American Society for Testing and Materials (ASTM) [35, 36] and the Deutches Institut for Normung e.V. (DIN) [37]. Although outdated, the ASTM has published a series of bibliographies on SEC [38-41] which is still useful. Updated comprehensive reviews on SEC are published biennially in the Fundamental Review issues of *Analytical Chemistry* [42-50].

Research papers on SEC and related techniques can be found in many analytical and polymer chemistry journals, including *Analytical Chemistry, Journal of Polymer Science, Journal of Applied Polymer Science, Polymer, Macromolecules, International Journal of Polymer Analysis and Characterization, Journal of Chromatography, Journal of Liquid Chromatography*, and *Chromatographia*. Finally, at the time of this writing, there are four major meetings that deal with polymer chromatography that are held on a regular basis: the International Symposium on Polymer Analysis and Characterization, International GPC Symposium, the ACS Symposium on SEC, the National Symposium of Polymer Analyis and Characterization (Japan), and the Bratislava International Conference on Macromolecules: Chromatography of Polymers and Related Substances (Slovakia). Information about these meetings can be found in published announcements or on appropriate Internet Web sites.

References

1. BARTH HG, MAYS JW (eds) (1991) Modern Methods of Polymer Characterization. Wiley, New York
2. COOPER AR (ed) (1989) Determination of Molecular Weight. Wiley, New York
3. EZRIN M (ed) (1973) Polymer Molecular Weight Measurements. Adv Chem Ser 125. American Chemical Society, Washington, DC
4. RABEK JF (1980) Experimental Methods in Polymer Chemistry. Wiley, New York
5. MORI S (1978) Anal Chem 50:424A
6. PORATH J, FLODIN P (1959) Nature 183:1657
7. MOORE JC (1964) J Polym Sci A2:835
8. GIDDINGS JC (1967) Anal Chem 39:1027
9. BARTH HG (1995) In: PROVDER T, BARTH HG, URBAN MW (eds) Chromatographic Characterization of Polymers. Adv Chem Ser 247. ACS Symposium Ser 352, Chap. 1. American Chemical Society, Washington, DC
10. YAU WW, KIRKLAND JJ, BLY DD (1979) Modern Size Exclusion Chromatography. Wiley, New York
11. KREMMER T, BOROSS L (1979) Gel Chromatography. Wiley, New York

12. DETERMANN H (1968) Gel Chromatography. Springer, Berlin
13. ALTGELT KH, SEGAL L (eds) (1971) Gel Permeation Chromatography. Dekker, New York
14. JANČA J (ed) (1984) Steric Exclusion Liquid Chromatography. Dekker, New York
15. HUNT BJ, HOLDING SB (eds) (1989) Size Exclusion Chromatography. Chapman and Hall, New York
16. DUBIN PL (ed) (1988) Aqueous Size Exclusion Chromatography. Elsevier, Amsterdam
17. WU C-S (ed) (1995) Handbook of Size Exclusion Chromatography. Dekker, New York
18. CAZES J (ed) (1979) Liquid Chromatography of Polymers and Related Materials. Dekker, New York
19. CAZES J, DELAMARE X (eds) (1980) Liquid Chromatography of Polymers and Related Materials II. Dekker, New York
20. CAZES J (ed) (1981) Liquid Chromatography of Polymers and Related Materials III. Dekker, New York
21. PROVDER T (ed) (1980) Size Exclusion Chromatography. ACS Symposium Ser 138. American Chemical Society, Washington, DC
22. PROVDER T (ed) (1984) Size Exclusion Chromatography, ACS Symposium Ser 245. American Chemical Society, Washington, DC
23. PROVDER T (ed) (1987) Detector and Data Analysis in Size Exclusion Chromatography. ACS Symposium Ser 352. American Chemical Society, Washington, DC
24. PROVDER T (ed) (1993) Chromatography of Polymers. ACS Symposium Ser 521. American Chemical Society, Washington, DC
25. PROVDER T, BARTH HG, URBAN MW (eds) (1995) Chromatographic Characterization of Polymers. Adv Chem Ser 247. American Chemical Society, Washington, DC
26. POTSCHKA M, DUBIN PL (eds) (1996) Strategies in Size Exclusion Chromatography, ACS Symposium Ser 635. American Chemical Society, Washington, DC
27. Waters Corporation, 34 Maple St., Milford, MA 01757
28. TUNG LH (ed) (1977) Fractionation of Synthetic Polymers. Dekker, New York
29. GLÖCKNER G (1987) Polymer Characterization by Liquid Chromatography. Elsevier, Amsterdam
30. GLÖCKNER G (1991) Gradient HPLC of Copolymers and Chromatographic Cross-Fractionation. Springer, Berlin
31. BALKE ST (1984) Quantitative Column Liquid Chromatography. Elsevier, Amsterdam
32. BELENKII BG, VILENCHIK LZ (1983) Modern Liquid Chromatography of Macromolecules. Elsevier, Amsterdam
33. FRANCUSKIEWICZ (1994) Polymer Fractionation. Springer, Berlin
34. PASCH H, TRATHNIGG B (1998) HPLC of Polymers. Springer, Berlin
35. ASTM D3536-76 (Revised 1986) Test Method for Molecular Weight Averages and Molecular Weight Distribution of Polystyrene by Liquid Exclusion Chromatography (1986-10)
36. ASTM D3593-80 (Revised 1986) Test Method for Molecular Weight Averages and Molecular Weight Distribution of Certain Polymers by Liquid Size Exclusion Chromatography Using Universal Calibration (1986-10)
37. DIN 55672-1 Gelpermeationschromatographie Teil 1: Tetrahydrofuran als Elutionsmittel (1995-02)
38. Bibliography on Liquid Exclusion Chromatography (1974) AMD 40, ASTM
39. Bibliography on Liquid Exclusion Chromatography 1972–1975 (1977) AMD 40-S1, ASTM
40. Bibliography on Liquid Exclusion Chromatography 1976–78 (1981) AMD 40-S2, ASTM
41. Bibliography on Size Exclusion Chromatography 1979–82 (1985) AMD 40-S3, ASTM.
42. HAGNAUER GL (1982) Anal Chem 54:265R
43. MAJORS RE, BARTH HG, LOCHMULLER CH (1984) Anal Chem 56:300R

44. BARTH HG, BARBER WE, LOCHMULLER CH, MAJORS RE, REGNIER FE (1986) Anal Chem 58:211R
45. BARTH HG, BARBER WE, LOCHMULLER CH, MAJORS RE, REGNIER FE (1988) Anal Chem 60:387R
46. BARTH HG, BOYES BE (1990) Anal Chem 62:381R
47. BARTH HG, BOYES BE (1992) Anal Chem 64:428R
48. BARTH HG, BOYES BE, JACKSON C (1994) Anal Chem 66:595R
49. BARTH HG, BOYES BE, JACKSON C (1996) Anal Chem 68:445R
50. BARTH HG, BOYES BE, JACKSON C (1998) Anal Chem 70:251R

2 Fundamental Concepts

2.1 Separation Mechanism of SEC

Molecular size, or more precisely, molecular hydrodynamic volume governs the separation process of SEC. That is, as a mixture of solutes of different size passes through a column packed with porous particles, the molecules that are too large to penetrate the pores of the packing elute first, as shown in Fig. 2.1. Smaller molecules, however, that can penetrate or diffuse into the pores, elute at a later time or elution volume. Thus a sample is separated or fractionated by molecular size, the profile of which describes the molecular weight distribution (MWD) or size distribution of the mixture. If the SEC system is calibrated with a series of solutes of known MW, as shown in Fig. 2.2, a relationship between log MW and elution volume is obtained. This relationship can then be used as a calibration curve to determine the MWD and MW averages of samples, as explained in Chap. 7.

It can easily be seen from Fig. 2.2 that high-MW solutes, which are too large to penetrate the pores of the packing, elute within the interstitial or void volume V_0 of the column, that is, the volume of mobile phase that is located between the packing particles. As the molecular size of the solute becomes smaller and

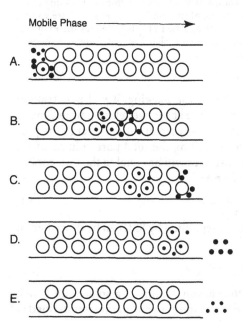

Mobile Phase

Fig. 2.1. Schematic of an SEC separation showing the separation of low (•) and high MW (•) polymers: *A* start of separation; *B* smaller molecules diffuse into porous particles, while larger molecules elute in the interstitial regions of the packed bed; *C* size separation is completed; *D* large molecules, which sample less column volume, elute first; *E* small molecules having access to both interstitial and pore volumes elute later

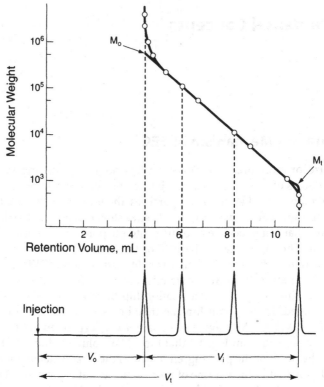

Fig. 2.2. A typical SEC calibration curve showing a plot of log MW vs. elution or retention volume of a series of polymer standards of known MW. The total permeation volume of the column V_t, is equal to the exclusion or void volume of the column, V_0, plus the pore volume of the packing within the column V_i. M_0 is the extrapolated MW that defines the exclusion limit of the column occurring at V_0, and M_t is the extrapolated MW that defines the total permeation limit of the column occurring at V_t

begins to approach the average pore size of the packing, it will penetrate or partition into the pores of the packing and elute at a longer elution time. Finally, when the molecular size of the solute is relatively small with respect to the pore size, it will freely diffuse into the pores sampling the total pore volume of the packing V_i. The elution volume of small solutes will be equal to the total mobile phase volume V_t of the packed SEC column

$$V_t = V_0 + V_i \tag{2.1}$$

The chromatographic behavior of solutes separated by SEC can be described by the general chromatographic equation

$$V_R = V_0 + K_{SEC} V_i \tag{2.2}$$

where V_R is the retention volume or the elution volume of a solute and K_{SEC} is the SEC distribution coefficient. K_{SEC} is a thermodynamic parameter defined as the ratio of the average concentration of the solute in the pore volume $\langle c \rangle_i$ to that in the interstitial volume $\langle c \rangle_0$

$$K_{SEC} = \langle c \rangle_i / \langle c \rangle_0 \tag{2.3}$$

Equations (2.2) and (2.3) demonstrate very well that K_{SEC} has defined limits of $0 \leq K_{SEC} \leq 1$. As a result, the product of the interstitial volume and the accessible pore volume, $K_{SEC}V_i$, governs the retention volume of a sample. When developing an SEC method, we must be sure that the separation mechanism is based entirely on molecular volume or size, that is, $0 < K_{SEC} < 1$. If $K_{SEC} > 1$, the separation is controlled by enthalpic interactions which depend on solute chemical composition and not necessarily on the MW (see Chap. 11). Furthermore, if $K_{SEC} = 0$, the sample will elute in the void volume; if $K_{SEC} = 1$, it will elute in the total column volume – in both cases separation will not occur.

Figure 2.3 shows the relationship between SEC and other forms of liquid chromatography. During interactive modes of chromatography, such as reversed-phase, normal-phase, or ion-exchange chromatography, solutes are retained on the packing via enthalpic interactions and elute after V_t. In fact, with these types of separations, V_t is designated as the elution time of an un-retained solute, or t_0. Furthermore, peaks eluting before t_0 are sometimes called "system peaks", and are usually ignored. In reality, however, the components eluting before t_0 are those that cannot completely penetrate into the pores of the packing because of size exclusion or electrostatic repulsive forces.

In principle, it is possible to use any HPLC column for SEC analysis by adjusting the mobile phase conditions to eliminate enthalpic interactions (albeit, pore

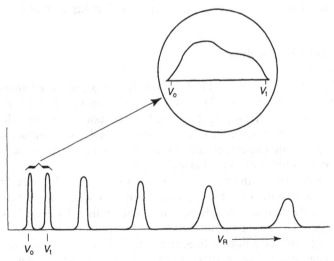

Fig. 2.3. Chromatogram showing both SEC and interactive modes of HPLC separation. Inset depicts the region where size separation occurs. V_0 is the interstitial volume, V_t is the total elution volume of the column ($V_0 + V_i$), and V_R is the elution or retention volume. (Reprinted from [43] by permission of Advanstar Communications Inc.)

size and volume should be optimized). Likewise, any SEC column can be converted into an interactive HPLC column by modifying the mobile phase in order to promote enthalpic interactions.

The elution volume of a component that takes into account both size exclusion (i.e., entropic) and enthalpic interactions can be formulated as:

$$V_R = V_0 + K_{SEC}V_i + K_{part}V_s + K_{ads}S \tag{2.4}$$

where K_{part} is the partition coefficient, V_s is the volume of the stationary phase, K_{ads} is the adsorption coefficient, and S is the surface area of the packing. To ensure that the separation mechanism is based only on size exclusion, the last two terms in Eq. (2.4) should be equal to zero.

In practice, a small amount of interaction between the packing and the solute molecules is sometimes observed and, therefore, K_{SEC} can be defined by [1]

$$K_{SEC} = K_D K_p \tag{2.5}$$

where K_D is the distribution coefficient for *pure* size exclusion and K_p is the distribution coefficient for solute-packing interaction effects. If solute molecules are separated solely by size exclusion and no enthalpic interactions are observed, then K_p is unity and retention volume for SEC is expressed by Eq. (2.2).

Since the introduction of SEC, there have been a number of separation mechanisms proposed to help explain its nature; these models include geometric considerations [2–12], restricted diffusion [13–16], separation by flow [17–22], and a thermodynamic model [23–29]. The mechanism of SEC separation has not yet been fully elucidated; however, the geometric approach and the thermodynamic model offer the clearest and practical explanations.

2.1.1 Geometric Models

The simplest models to understand intuitively are based on geometric considerations, which are shown in Fig. 2.4. The models are based on either conical (Fig. 2.4A) [5] or cylindrical (Fig. 2.4B) [2, 3, 10, 11] pore openings although, in reality, pore structures are much more complex. Molecular conformation or shape also plays an important role in developing geometric models; here, however, we will only consider spherical shapes.

In the conical model, the pore narrows with increasing distance. Spherical molecules penetrate the pore to a certain depth depending on the ratio of the molecule-to-pore diameter. A spherical molecule with a diameter larger than the pore (Fig. 2.4A, No. 1) cannot penetrate and is totally excluded from the pore. The large molecule moves directly through the column and appears first in the chromatogram. The smallest molecule (No. 3) can reach deep into the pore and penetrate almost the entire pore volume. It is retained the most and elutes last. An intermediate size molecule (No. 2) can penetrate into the pore until it reaches a diameter that is comparable to the size of the molecule. Thus

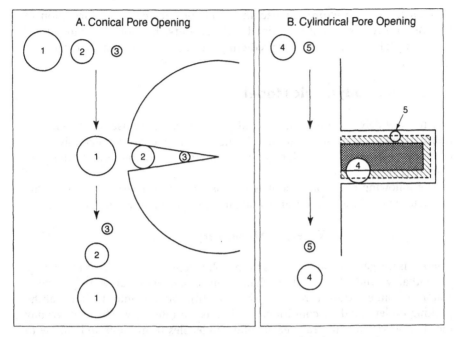

Fig. 2.4. Geometric models: *A* conical-pore opening; *B* cylindrical-pore opening

these three molecules elute in the order of decreasing molecular diameters: i.e. 1, 2, and 3.

The cylindrical pore model assumes the same diameter throughout the inside of the pore (Fig. 2.4B). The pore volume available for spherical molecules, however, is not equal to the entire pore volume. The accessible pore volume is limited to the radius of the molecule from the pore wall. For example, the accessible pore for a larger molecule (No. 4) (the cross-shaded area surrounded by a solid line) is smaller than the accessible pore for a smaller molecule (No. 5) (the diagonally shaded area surrounded by a dotted line), and elutes earlier from the column. The large molecule (No. 4) samples less pore volume and elutes from the column first, followed by the smaller molecule (No. 5).

Geometric models are, in general, better defined for high-performance SEC packings as compared to softer gels because HPLC packing materials are rigid and swell less; pore shape and volume do not change much with the nature of the mobile phase. Conical and cylindrical representations for soft gels, such as dextran, however, are difficult to employ, and a molecular sieve concept is preferred. Dextran-based gels, for example, consist of a three-dimensional, swollen network. The degree of cross-linking and the composition of the mobile phase control the dimensions of the gel pore: the less cross-linking, the greater the swelling, resulting in large gel pores. In this case, smaller molecules penetrate into the network, sample a greater pore volume and elute last.

Because pore structures of packings and the molecular conformation of polymer molecules cannot be defined with certainty, the geometric models, for the most part, are not useful for predicting SEC chromatographic behavior.

2.1.2 Thermodynamic Model

The most widely accepted theory, and perhaps the most useful, is based on thermodynamic considerations, that is, the establishment of a thermodynamic equilibrium between the polymer in the interstitial volume and in the pore volume [23].

The following equation describes the distribution process, in terms of the distribution coefficient K, for any solute in equilibrium between two phases

$$K = \exp(-\Delta H^\circ/RT) \exp(\Delta S^\circ/R) \tag{2.6}$$

where R is the gas constant and T is the absolute temperature. ΔH° is the change in enthalpy and ΔS° is the change in conformational entropy when 1 mole of solute is transferred from the interstitial volume into the pores of the packing under standard conditions. In the case of interactive chromatography the sign of ΔH° for the transfer of solute molecules from the mobile phase to the pore volume (or stationary phase) is usually negative. At ideal SEC conditions, however, ΔH° is equal to zero; that is, there are no enthalpic interactions between the solute and packing. Thus $K = K_{SEC}$ and Eq. (2.6) becomes

$$K_{SEC} = \exp(\Delta S^\circ/R) \tag{2.7}$$

Since the conformational degrees of freedom of a macromolecule are more restricted inside the pores of the packing, as compared to being in the interstitial volume, the conformational entropy of a chain decreases during permeation into the pores. The driving force behind ΔS° arises from the concentration gradient of the solute developed between the interstitial volume and the pore volume. The loss in conformational entropy when a polymer molecule transfers from the mobile phase to within a pore governs SEC separation. The restraint of conformational freedom of a flexible polymer chain in the confines of a pore causes a decrease in entropy. As a result, the sign of ΔS° is negative and K_{SEC} values range from zero to unity. Figure 2.5 is a schematic representation of the influence of molecular size on ΔS° and K_{SEC}. In panel A, the solute can freely sample the entire pore volume; as a result, there is no measurable change in ΔS°. In panel B, the macromolecules have approached the pore opening dimensions and occupy a finite volume of the pore; in this case ΔS° is negative and K_{SEC} decreases. Finally, panel C depicts an excluded chain, in which ΔS° takes on a large negative value, and K_{SEC} approaches zero.

Unlike all other forms of separations, the SEC distribution coefficient is independent of temperature and depends only on the size and shape of the macromolecule with respect to the average pore size (and shape) of the

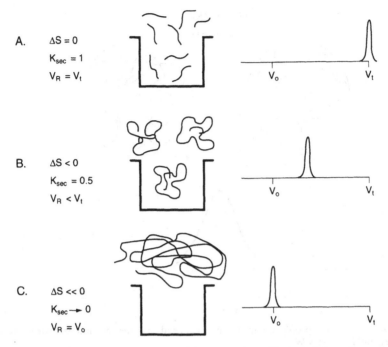

Fig. 2.5. Thermodynamic representation of the SEC process. Elution of: *A* totally permeated molecules; *B* partially excluded molecules; *C* totally excluded molecules

packing. Thus, in principle, the elution volume, or more precisely the first moment of a polydisperse polymer peak, is independent of column temperature. There are, however, secondary effects that will be affected by temperature, such as the change in pore size if nonrigid packings are used. Furthermore, molecular size may also change with temperature, although this effect is relatively small and can be either negative or positive depending on the nature of the mobile phase. More important, however, is the effect of temperature on column efficiency; increased column temperature decreases peak broadening, which results in more accurate MW determinations (see Sect. 2.3.3). Obviously, elevated temperatures may be necessary to ensure sample solubility. In practice, however, it is best to thermostat the column, which also ensures detector baseline stability, especially if a differential refractometer is used.

2.2 Hydrodynamic Volume Concept

Since the separation in SEC is regulated by the size of solute molecules in solution, solutes of similar hydrodynamic volume but different molecular weight will elute at the same retention volume. This concept is demonstrated in Fig. 2.6 which shows a highly branched chain eluting at the same retention volume as a

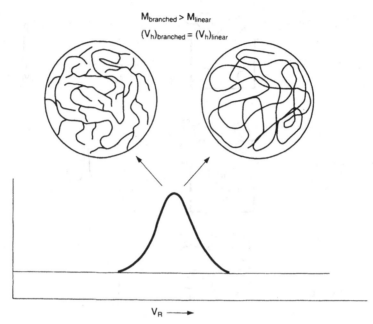

Fig. 2.6. SEC elution profile of linear and branched samples of similar hydrodynamic volumes, but different MWs

linear, lower-MW polymer of similar hydrodynamic volume. Likewise, solutes of similar MW but different molecular conformation will elute at different retention volumes. Thus a rod-shaped molecule, which will sweep out a larger volume than a flexible-coil molecule having the same MW, will elute earlier. A spherical molecule having the same MW as the flexible-coil molecule will have a more compact size and will elute later.

The effect of molecular shape on SEC calibration is shown in Fig. 2.7. As an example, the retention volume of a globular protein will be increased after it is denatured to form a more extended conformation.

The size of a polymer coil in solution can be extended or contracted depending on the solvent and temperature. Because of the fact that polymer–solvent interactions influence the size of a molecule in solution, "molecular size" must be defined in terms of solvent and temperature conditions. Furthermore, most polymers are flexible coils; consequently, their size must be expressed in terms of statistical parameters [30]. As illustrated in Fig. 2.8, the most commonly used size parameter is the radius of gyration R_g, defined as the root-mean-square distance R_i of the elements of the chain of mass m_i from its center of gravity

$$\langle R_g^2 \rangle^{1/2} = (\textstyle\sum_i m_i R_i^2 / \sum_i m_i)^{1/2} \tag{2.8}$$

The radius of gyration can be determined experimentally from light scattering measurements, provided that the MW is greater than about 10^5 [31], or by intrinsic viscosity measurements if the chains are linear, as explained below.

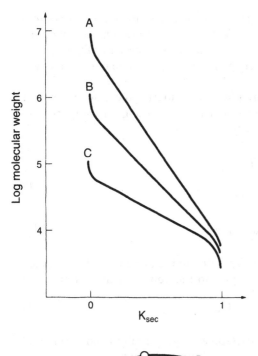

Fig. 2.7. SEC calibration curves of differently shaped molecules: A spherical; B random coil; C rod-like. (Yau WW, Kirkland JJ, Bly DD (1979) Modern Size Exclusion Chromatography, Wiley, New York, (© 1979, John Wiley & Sons). Reprinted by permission of John Wiley & Sons, Inc.

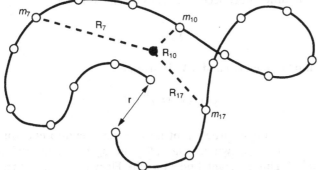

Fig. 2.8. Schematic of a random-coil chain consisting of 20 mass elements m_i, where R is the distance of the mass elements to the center of gravity •, and r is the end-to-end distance of the chain. (Adapted from [43] by permission of Advanstar Communications Inc.)

In order to relate MW to molecular size, the concept of intrinsic viscosity, defined as $[\eta]$, is often used. The intrinsic viscosity of a polymer in solution can be easily determined and is related to the MW, size and shape of the macromolecule in solution. When a polymer is dissolved in a solvent, the solution viscosity or its resistance to flow increases due to the presence of solvated polymer chains. In order to determine intrinsic viscosity accurately, measurements are carried out in dilute solutions and are extrapolated to infinite dilution to eliminate polymer–polymer interaction [32].

Intrinsic viscosity is relatively easy to measure using simple apparatus and is a fundamental molecular size parameter in a given solvent and temperature. Because intrinsic viscosity plays a major role with respect to SEC universal calibration, as will be shown in later chapters, it is important to understand how measurements are made.

The *relative* viscosity η_r is measured first by comparing the time t it takes for a polymer solution of given concentration to flow through a capillary with respect to the flow time t_0 of the solvent

$$\eta_r = t/t_0 \tag{2.9}$$

The *specific* viscosity η_{sp} is then calculated as

$$\eta_{sp} = \eta_r - 1 \tag{2.10}$$

which measures the increase in viscosity owing to the presence of the polymer. This value is then normalized with respect to the polymer concentration c to give the *reduced* viscosity

$$\eta_{red} = \eta_{sp}/c \tag{2.11}$$

The *intrinsic* viscosity $[\eta]$ is then extrapolated graphically from a series of measurements at different concentrations to infinite dilution

$$[\eta] = (\eta_{sp}/c)_{c \to 0} \tag{2.12}$$

The relationship between $[\eta]$ and MW is given by

$$[\eta] = \phi_o \langle r^2 \rangle^{3/2}/M \tag{2.13}$$

where $\langle r^2 \rangle^{1/2}$ is the unperturbed root-mean-square end-to-end distance of the polymer chain, as shown in Fig. 2.8, and ϕ_o is a universal constant, 2.1×10^{23} (mol)$^{-1}$. This constant, known as the Flory constant, is theoretically the same for all linear, flexible chains that are in an unperturbed state [32]. In other words, the chains approach random-coil configurations under a given set of solvent and temperature conditions, also known as theta conditions.

Equation (2.13), referred to as the Flory–Fox equation [33], relates molecular volume $\langle r^2 \rangle^{3/2}$ and MW to intrinsic viscosity. Rather than using the root-mean-square end-to-end distance in Eq. (2.13), the radius of gyration is commonly accepted, in which $\langle r^2 \rangle = 6 \langle R_g^2 \rangle$ [30]. Substituting this term into Eq. (2.13)

$$[\eta] = \phi_o 6^{3/2} \langle R_g^2 \rangle^{3/2}/M \tag{2.14}$$

Replacing the molecular hydrodynamic volume V_h for $\langle R_g^2 \rangle^{3/2}$ and rearranging, we obtain

$$[\eta]M = \phi_o 6^{3/2} V_h \tag{2.15}$$

This relationship tells us that $[\eta] M$ is related to the molecular hydrodynamic volume of a polymer. V_h is a key parameter in SEC separation and, as will be shown in Chap. 7, is needed to understand the concept of SEC universal calibration.

Another useful equation that relates intrinsic viscosity to molecular weight is the Mark–Houwink equation

$$[\eta] = KM^a \qquad (2.16)$$

in which a and K are coefficients for a given polymer dissolved in a specified solvent and temperature. The exponent a can be considered to be a conformational parameter of the macromolecule, ranging from zero for a spherical conformation, that is, there is no intrinsic viscosity dependency on MW, to a value of two for a rigid-rod conformation. For flexible macromolecules, $0.5 < a < 0.8$. Values < 0.5 indicate branched polymer structures, while values > 0.8 indicate a more extended molecular conformation. The Mark–Houwink equation is useful for calculating the MW of a polymer from measured values of $[\eta]$ provided that published coefficients are available. In addition, if narrow-MW fractions of a polymer are available, a can be determined from the slope of a plot of log $[\eta]$ vs. log M. Since the magnitude of a reflects polymer conformation, the extent of long-chain branching also can be estimated (Chap. 12).

2.3 SEC Parameters and Definitions

SEC parameters of interest relate to column performance, resolution and the characteristics of the calibration curve, all of which are covered in this section. Also treated are standard chromatographic terms and definitions, including specific terminology related to SEC resolution. Fore more in-depth treatment of the general principles and concepts of HPLC, standard texts on the subject should be consulted [34, 35]. The significance of MW averages, their calculations and relationship to elution volume are discussed in Chaps. 6 and 7.

2.3.1 Column and Calibration Curve Characteristics

In SEC the interest is in determining the retention volume at equally spaced intervals across the peak profile with respect to concentration. Although either elution time or volume may be used, it is better practice to use retention volume V_R because it is independent on the flow rate F

$$V_R = t_R F \qquad (2.17)$$

where t_R is defined as the time (sec or min) measured from the start of injection to the location of the portion of the peak of interest.

The total SEC column volume V_c consists of the interstitial or void volume V_o, pore volume of the packing V_i, and the volume occupied by the gel matrix or solid support V_g

$$V_c = V_o + V_i + V_g \tag{2.18}$$

Typically, V_o is approximately 35 % of V_c, and the remaining 65 % is divided between V_g and V_i; the percentage of each depends on the nature and pore size of the packing. For example, V_g can range from about 20 % for soft gels to 30 % for more rigid structures, as in the case of silica particles. The volume of the solid support plays no direct role in SEC separations; however, as discussed below, the V_i/V_o ratio (or slope of the calibration curve) is a significant factor in terms of column resolution.

As shown in Fig. 2.2, polymers elute between V_0 and V_t, where V_t is the total mobile phase volume described in Eq. (2.1). Polymers that cannot enter the pores of the packing elute within the interstitial volume (also referred to as the void or exclusion volume) V_o of the column. Those molecules that can occupy the interstitial *and* total pore volume of the packing elute at V_t, the total permeation region of the column. The extrapolated MW values occurring at V_o and V_t are defined as the exclusion molecular weight M_o and the total permeation molecular weight M_t, respectively. In practice, M_o and M_t define the separation range of a given column. For large pore size packings, it is sometimes difficult to find a polymer of sufficiently high MW that is excluded. In this case, V_o can be estimated as 35 % of the total column volume. The pore volume of the column V_i is perhaps the most critical parameter in SEC, for within V_i separation takes place.

Another important characteristic of an SEC column is the slope of the calibration curve. Even though there may be some curvature associated with log MW vs. log V_R plots, modern SEC columns are in general fairly linear and the following equation is a good approximation [23]:

$$D_2 = (\log M_o - \log M_t)/V_i = \log(M_o/M_t)/V_i \tag{2.19}$$

In comparison to enthalpic modes of separation, D_2 can be related to column selectivity. That is, D_2 defines how well a column can differentiate peak *maxima* of polymers of different MW values. For optimal separation of a given sample, the SEC column must have an adequate separation range to cover the entire MW limits of the sample and should have a minimum D_2 value for greatest separation. For maximum selectivity, i.e. minimum slope, the pore volume should be as large as possible. This effect can be best explained using Fig. 2.9 [23]. Here we see calibration curves from a single column (Fig. 2.9A) having $V_i = 2.5$ ml, and two columns (Fig. 2.9B) where $V_i = 5$ ml, with respective *reciprocal* slopes of about 0.8 and 1.6 ml per decade of MW. The two-column set has twice the selectivity or 40 % higher resolution (see Eq. (2.28)). As discussed in the following section, the third requirement for optimum resolution is column efficiency.

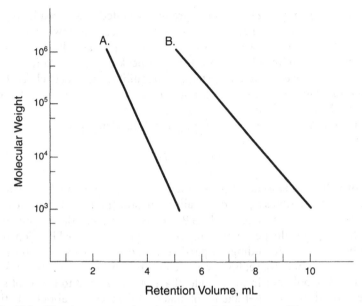

Fig. 2.9. SEC calibration curves using single *(A)* and dual columns *(B)*

Since the retention volume of a polymer V_R is dependent on column dimensions, it is sometimes advantageous to use the distribution coefficient K_{SEC} when comparing chromatographic behavior of polymers with different columns or for fundamental studies

$$K_{SEC} = (V_R - V_o)/V_i \tag{2.20}$$

2.3.2 Peak Broadening

As a polymer elutes through a column, it will be fractionated on the basis of molecular size by the SEC process. The injected sample will begin to spread out or broaden as it travels through the column because of the presence of different molecular size polymers. In this case, the degree of peak broadening, i.e. resolution, will depend on the polydispersity of the sample, the pore volume, and the slope of the calibration curve. Superimposed on this separation, however, are other factors that contribute to peak broadening, which interfere with the separation process. One important source of unwanted peak broadening is caused by *extra-column* volume from the sample injection, detector cell, and interconnecting tubing. These volume contributions should be kept to a minimum with respect to V_t.

This type of peak broadening, *unrelated to sample polydispersity*, can be defined in terms of a peak standard deviation σ or, more precisely, peak variance in units of length σ_L^2, time σ_t^2, or volume σ_V^2. Because variances are additive, the overall contribution to peak broadening is

$$(\sigma_L^2)_{total} = (\sigma_L^2)_{inj} + (\sigma_L^2)_{det} + (\sigma_L^2)_{tubing} + (\sigma_L^2)_{column} \tag{2.21}$$

The last term in Eq. (2.21) is a measure of unwanted peak broadening that occurs *within* the SEC column. This term takes into account flow inhomogeneities in the packed column, whereby the injected sample can take different flow pathways resulting in broadened and distorted peak profiles. Also included in $(\sigma_L^2)_{column}$ is resistance to mass transfer, a nonequilibrium effect related to the finite time or distance it takes for a polymer to diffuse from the interstitial volume into the pore volume, and then back out again.

A useful equation that relates $(\sigma_L^2)_{column}$ to column length is

$$(\sigma_L^2)_{column} = HL \qquad (2.22)$$

which shows that the amount of peak broadening is proportional to the length of the SEC column L times a proportionality constant H referred to as the *height equivalent to a theoretical plate* (HETP). Thus, as expected, the longer the column, the greater is the peak spreading. The magnitude of HETP is dependent on a number of factors including the particle diameter of the packing, the linear velocity of the mobile phase, how well a column is packed, and the diffusion coefficient of the polymer in the mobile phase. As compared to HPLC of small molecules, the diffusion coefficients of macromolecules are about 10^2 times smaller, which will be reflected in increased peak broadening.

Related to HETP is the number of theoretical plates N generated in a column

$$N = L^2/(\sigma_L^2)_{column} \qquad (2.23)$$

or

$$N = L/H \qquad (2.24)$$

Converting from units of column length to retention volume, we obtain

$$N = V_R^2/(\sigma_v^2)_{column} \qquad (2.25)$$

Assuming a Gaussian peak shape, the standard deviation σ is equal to $w/4$, where w is the peak width in volume units, that is, $w = w_t F$. Thus N can be calculated directly from a chromatogram using

$$N = 16(V_R/w)^2 = 5.54\,(V_R/w_{1/2})^2 \qquad (2.26)$$

in which w and $w_{1/2}$ are the peak widths at baseline and half height, respectively, as defined in Fig. 2.10. The standard deviation can also be taken as one-half the peak width at the inflection point, which is located at 0.607 of the height from the baseline. Peak width at the baseline is measured by extension of the tangents at the inflection points of the peak to the baseline.

It is important to note that the plate height theory described above assumes a Gaussian peak distribution, which is not usually the case in most chromatographic processes. Furthermore, SEC column efficiencies are typically reported using low-MW solutes, which give significantly higher values than for polymers. One practical reason for this convention is that the SEC peak width

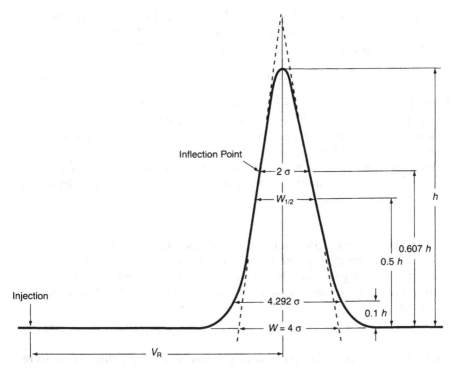

Fig. 2.10. Idealized Gaussian-shaped chromatographic peak indicating different methods of measuring peak standard deviation σ

depends on the polydispersity and MW of the polymer, and most standards have finite polydispersities.

The value of N for SEC columns is usually measured by injecting 20 μl of a 0.1% ethylbenzene solution [36]. The value of N varies with the nature of the test sample, injection volume and other experimental conditions. The value of N is usually reported as the number of plates per meter or total column length. To obtain the former value, N calculated from Eq. (2.26) is multiplied by $100/L$, where L is the column length in cm. The larger the number, the lower the amount of chromatographic peak broadening. This, as will be demonstrated, results in higher resolution and increased MW accuracy.

2.3.3 Column Resolution

As in the case of all chromatographic separations, peak resolution is a critical parameter in SEC that incorporates column efficiency, N or H, and column selectivity, defined as the slope D_2 of the calibration curve (Eq. (2.19)). (In enthalpic modes of chromatography, the selectivity factor $\alpha = k_2'/k_1'$ is taken into account instead of D_2, where k_1' and k_2' are the capacity factors of the first and second peak, respectively; $k' = (t_R - t_0)/t_0$, in which t_R and t_0 are the corresponding retention times of the sample and unretained peaks.)

The resolution R_S of two adjacent peaks 1 and 2 is defined as

$$R_S = 2(V_{R2} - V_{R1})/(w_1 + w_2) \tag{2.27}$$

When $R_S = 1$, the two peaks are reasonably well separated with only 2% overlap of one peak on the other if heights of the two peaks are nearly equal. Larger values of R_S mean better separation. At $R_S = 1.25$, complete separation can be assumed and $R_S = 0.5$ means no separation.

The effect of column length L on resolution can be easily understood by referring to Eq. (2.27). Since $V_{R2} - V_{R1}$ is proportional to L, and from Eq. (2.22) σ_L scales with $L^{1/2}$, then

$$R_S \propto L^{1/2} \tag{2.28}$$

Thus, if the column length is doubled, R_S will increase by $\sqrt{2}$, or 40%.

In SEC a more appropriate resolution expression is needed to take into account that the separation is based on molecular size. Assuming a linear calibration curve, the resolution for a SEC column can be defined as a specific resolution [37]

$$R_{SP} = 2(V_{R2} - V_{R1})/(w_1 + w_2) \log_{10}(M_2/M_1) \tag{2.29}$$

This relationship describes the resolving power of an SEC column in terms of how well it can separate two polymer samples that differ by a decade of MW. In practice, R_{SP} is determined by injecting two calibration standards differing by a factor of about 10 in MW and the value of R_{SP} should be equal to or greater than 1.3 [38]. Alternatively, the two calibration standards should be approximately within the elution range of the half-height of the main peak of the sample analyzed and R_S (Eq. (2.27)) should be at least 2.5 [36].

Another form of Eq. (2.29) can be derived from the relationship of MW to the slope D_2 and intercept D_1 of the calibration curve [23]

$$M = D_1 e^{-D_2 V_R} \tag{2.30}$$

Combining this expression with Eq. (2.27), we obtain [23]

$$R'_S = \ln(M_2/M_2)/2D_2(\sigma_1 + \sigma_2) \tag{2.31}$$

where ln is the natural logarithm. Assuming that the standard deviations of peaks 1 and 2 are similar,

$$R'_S = \ln(M_2/M_2)/4\sigma D_2 \tag{2.32}$$

This interesting result states that SEC column resolution is directly linked to the product of the slope of the calibration curve and column efficiency: for optimum resolution we need both minimum slope and peak broadening. SEC columns of high plate numbers may not necessarily provide adequate resolution unless the slope of the calibration curve is sufficiently low.

The goal in SEC is to obtain accurate MW data; thus column performance needs to be related to MW accuracy. Several investigators [23, 39–42], who took into account the influence of peak broadening, have developed this connection. The relationship between the true and experimentally derived MW can be expressed as

$$MW_{true} = (X)MW_{exp} \tag{2.33}$$

in which X is a correction factor that depends on the mathematical model used. Yau and colleagues [23] developed the following expression that represents the relative MW error

$$(MW_{exp} - MW_{true})/MW_{true} = (1/X) - 1 \tag{2.34}$$

Because of peak broadening, the high-MW end of the MWD will broaden into the early eluting region; likewise, the low-MW end of the MWD will broaden out and elute into the later eluting region. The net result is an overestimation of M_w and an under estimation of M_n. The MW errors incurred for M_n and M_w, designated by $M*$, are

$$M_n^* = e^{-(1/2)(\sigma D_2)^2} - 1 \tag{2.35}$$

$$M_w^* = e^{(1/2)(\sigma D_2)^2} - 1 \tag{2.36}$$

$M*$ can be minimized if the exponential term approaches zero.

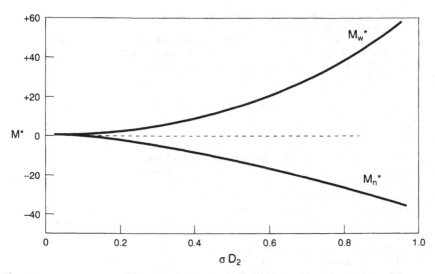

Fig. 2.11. MW error $M*$ of the computed M_n or M_w as a function of the product of chromatographic band broadening σ and the slope of the calibration curve D_2. (Adapted from J Chromatogr, Vol 125, YAU WW, KIRKLAND JJ, BLY DD, STOKLOSA HJ, "Effect of Column Performance on the Accuracy of Molecular Weights Obtained from Size Exclusion Chromatography", pp 219–230 (1976) by permission of Elsevier Science.)

Figure 2.11 is a graphical representation of Eqs. (2.35) and (2.36) which is quite useful for calculating errors associated with symmetrical peak broadening. The influence of peak skewing is not discussed here but can be found elsewhere [23, 39–42]. The slope D_2 is determined from Eq. (2.19) and σ_v is approximated from $(w/4) \cdot F$. For example, a column with a σD_2 value of 0.2 gives an acceptable MW error of 2%, while a column with a σD_2 value of 0.6 gives an unacceptable MW error of 20%. Since the slope of the calibration curve is inversely proportional to V_0 or column length ($D_2 \propto L^{-1}$) and σ is proportional to $L^{1/2}$ (see Eq. (2.28)) then σD_2 scales with $L^{-1/2}$. Using the above example, the column with a σD_2 value of 0.6 can be reduced to 0.2 by increasing the column length by a factor of nine. The disadvantage of this approach is that analysis time is compromised. A better way would be to start with a high plate-count column of the correct separation range and maximum pore volume. (It is important to emphasize that the above peak broadening errors do not include errors introduced by flow-rate variability, baseline drift and selection, and MW accuracy of the calibration standards (see Chap. 5).)

References

1. DAWKINS JV, HEMMING M (1975) Makromol Chem 176:1777, 1795, 1815
2. CASSASA EF (1967) J Polym Sci, Part B 5:773
3. CASSASA EF (1971) J Phys Chem 75:3929
4. GIDDINGS JC, KUCERA E, RUSSEL CP, MYERS MN (1968) J Phys Chem 72:4397
5. PORATH J (1963) J Pure Appl Chem 6:233
6. SQUIRE PG (1964) Arch Biochem Biophys 107:471
7. LAURENT TC, KILLANDER J (1964) J Chromatogr 14:317
8. OGSTON AG (1958) Trans Faraday Soc 54:1754
9. VAN KREVELD, VAN DEN HOED N (1973) J Chromatogr 83:111
10. YAU WW, MALONE CP (1971) Polym Prepr (Am Chem Soc, Div Polym Chem) 12:797
11. KUBIN M, VOZKA S (1980) J Polym Sci, Polym Symp 68:209
12. HAGER D (1980) J Chromatogr 187:285
13. ACKERS GK (1964) Biochem 3:723
14. SMITH WB, KOLLMANSBERGER (1965) J Phys Chem 69:4157
15. YAU WW, MALONE CP (1967) J Polym Sci., Part B 5:663
16. DIMARZIO EZ, GUTTMAN CM (1969) J Polym Sci, Part B 7:267
17. DIMARZIO EZ, GUTTMAN CM (1979) Macromolecules 3:131
18. GUTTMAN CM, DIMARZIO EZ (1970) Macromolecules 3:681
19. VERHOFF HF, SYLVESTER ND (1970) Macromol Sci – Chem A4:979
20. AUBERT JH, TIRRELL M (1980) Sep Sci Technol 15:123
21. AUBERT JH, TIRRELL M (1980) Rheol Acta 19:452
22. CHENG WJ (1986) J Chromatogr 362:309
23. YAU WW, KIRKLAND JJ, BLY DD (1979) Modern Size Exclusion Chromatography. Wiley, New York
24. YAU WW, MALONE CP, SUCHAN HL (1970) Sep Sci 5:259
25. YAU WW, SUCHAN HL, MALONE CP (1968) J Polym Sci Part A-2 6:1349
26. YAU WW, MALONE CP, FLEMING SW (1968) J Polym Sci Part B 6:803
27. CHANG TL (1968) Anal Chim Acta 42:51
28. HOAGLAND DA (1996) ACS Symp Ser 635:173
29. DAWKINS JV (1976) J Polym Sci, Polym Phys Ed 14:569
30. TANFORD C (1961) Physical Chemistry of Macromolecules. Wiley, New York

31. KRATOCHVIL P (1987) Classical Light Scattering from Polymer Solutions. Elsevier, Amsterdam
32. MAYS JW, HADJICHRISTIDIS N (1991) In: BARTH HG, MAYS JW (eds) Modern Methods of Polymer Characterization, Chapter 7. Wiley, New York
33. FLORY PJ, FOX TG (1951) J Am Chem Soc 73:1904
34. SNYDER LR, KIRKLAND JJ (1979) Introduction to Modern Liquid Chromatography. Wiley, New York
35. KATZ E, EKSTEEN R, SCHOENMAKERS P, MILLER N (1998) Handbook of HPLC. Dekker, New York
36. DIN 55672-1 Gelpermeationschromatographie (GPC) Teil 1: Tetrahydrofuran (THF) as Elutionsmittel (1995-02)
37. BLY DD (1968) J Polym Sci Part C 21:13
38. ASTM D3536-76 (Reapproved 1986) Test Method for Molecular Weight Averages and Molecular Weight Distribution of Polystyrene by Liquid Exclusion Chromatography (Gel Permeation Chromatography – GPC) (1986-10)
39. YAU WW, STOKLOSA HJ, BLY DD (1977) J Appl Polym Sci 21:1911
40. HAMIELIC AE, RAY WH (1969) J Appl Polym Sci 13:1319
41. PROVDER T, ROSEN EM (1970) Sep Sci 5:437
42. YAU WW (1977) Anal Chem 49:395
43. BARTH HG (1984) LC Magazine 2:24 (Jan)
44. YAU WW, KIRKLAND JJ, BLY DD, STOKLOSA HJ (1976) J Chromatogr 125:219

3 Instrumentation

3.1 SEC Systems Used at Room Temperature

SEC measurements can be carried out using an HPLC setup which consists of a solvent delivery system, a separation system, and a detection system. A solvent delivery system is an assembly of a solvent reservoir, an on-line degasser, and a solvent delivery pump. A pulse damper may be connected after the pump. A separation system consists of a sample injector and SEC columns with or without a thermostated column oven. A detection system is composed of a detector, a strip-chart recorder and a waste reservoir. A data handling system to analyze SEC chromatograms and to calculate molecular weight (MW) averages of samples can be used in place of a strip-chart recorder. An autosampler (an autoinjector) which is used to inject a sample solution into the separation system at a specified interval can also be used with the HPLC apparatus. A schematic diagram of a typical SEC assembly is shown in Fig. 3.1.

Glass bottles of 0.5–2 l capacity are used as containers to act as a solvent reservoir and a waste reservoir. Plastic bottles should not be used, (1) because the solvent can be contaminated by additives in the plastic bottle and (2) the bottle may be attacked by the solvent. Solvent in the reservoir is delivered through poly-(tetrafluoroethylene) (PTFE) tubing which has a sintered stainless steel filter attached to the end to prevent dust or small particles being delivered to the pump. The waste reservoir is also connected with PTFE tubing to the outlet of the detector.

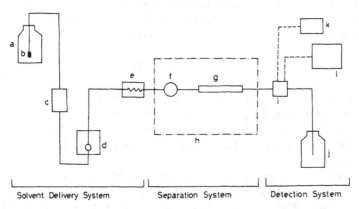

Fig. 3.1. Schematic diagram of a typical SEC assembly. *a* solvent reservoir, *b* filter, *c* in-line degasser, *d* pump, *e* pulse damper, *f* sample injector, *g* column, *h* column oven, *i* detector, *j* waste reservoir, *k* chart recorder, *l* data-handling system

The HPLC setup can be purchased as a completely integrated apparatus or are in a modular form. If operation at higher temperatures is not required, or if a specific application is needed, simple modular equipment may be adequate, as hardware requirements change somewhat in the course of time, e.g. smaller columns and a wider range of detectors are now available. Recently an in-line degasser has been made commercially available. This equipment is connected between the solvent reservoir and the pump and is used to remove dissolved air in the mobile phase, so that pre-degassing of the mobile phase is not required. PTFE tubing of 5 m × 2 mm i.d. (a volume of 15 ml) is installed in the vacuum chamber and continuous degassing is performed by running the solvent into the PTFE tube.

In order to obtain precise and accurate data needed for SEC, somewhat different equipment is required from that used in conventional HPLC. There are several requirements for the SEC apparatus. First, it is necessary to use a high-quality pump. Higher constancy of flow rate and good reproducibility are the most important factors for a solvent-delivery pump. Second, a loop valve must be used to inject the sample solution, and, third, the column temperature must be maintained constant. Keeping a constant temperature difference between the column and the solvent in the inlet of the solvent-delivery pump is recommended so that the mass velocity stays constant irrespective of the change in room temperature.

Besides these requirements, particular attention must be paid to the connections between the injector and the columns, and between the columns and the detector. The extra-column band broadening effects which occur outside the column, in addition to the inherent band broadening within the chromatographic column, should be kept to a minimum.

Extra-column dispersion expressed as as variance σ^2 is

$$\sigma^2 = \sigma^2_{inj} + \sigma^2_{tube} + \sigma^2_{cell} \tag{3.1}$$

where σ^2_{inj} is the variance related to the sample injection volume, σ^2_{tube} is the variance contributed by the connecting tubing, and σ^2_{cell} is the contribution from the detector cell volume. σ^2_{tube} is expressed as

$$\sigma^2_{tube} = (\pi r^4 lF)/(24 D_m) \tag{3.2}$$

where r and l are the radius and the length of the connecting tubing, F is the volumetric flow rate, and D_m is the solvent diffusion coefficient. Therefore, the connecting tubing should be as short and narrow as possible. These extra-column band broadening effects can be serious especially in microbore chromatography.

Integrated SEC setups are commercially available: Waters 150 C ALC/GPC from Waters Incorporated, Shodex GPC System-11 and System-21 from Showa Denko Co., Toso HLC-8020 from Toso Co., and PL-GPC 110 and PL-GPC 210 from Polymer Laboratories. The Waters 150 C ALC/GPC and PL-GPC 210 can also be used for high-temperature SEC (see next section). These instruments incorporate an injector, columns, a column oven and a refractive index detector. Although the Shodex GPC System-11 and System-21, Toso HLC-8020 and PL-GPC110 have a column oven and can control the column temperature up to 99 °C (for PL-GPC110 up to 110 °C), they can only be used for samples which are soluble in a solvent used as a mobile phase at room temperature, because the connecting tubing between the column and the detector and between the detector and the waste reservoir are not maintained at the same column temperature. The refractive index detector is also thermostated at 35 °C or up to 60 °C. The Shodex GPC System-11 and System-21 have facilities to maintain the inlet temperature from the outlet of the solvent reservoir to the sample injector constant through the pump. The inlet temperature can be adjusted to the same temperature as the column oven or to constant difference between the inlet and the column temperature. These systems also have a solvent degasser.

3.2 High-Temperature SEC Systems

High-temperature SEC (HTSEC) systems are used for polymers which are not soluble in any solvents at room temperature, or whose solutions are too viscous at lower temperature. Polyolefins, such as polyethylene and polypropylene, polyether ether ketone (PEEK), and polyphenylene sulfide (PPS) require high temperatures for dissolution in mobile phases. HTSEC systems require good temperature control from the solvent reservoir to the waste reservoir in addition to the requirements for conventional SEC systems. When an HTSEC system is considered, it usually means that the column temperature is over 100 °C. The temperature in the HTSEC apparatus should be kept over 100 °C not only in the injector columns, and detector(s), but in the connecting tubing between the injector and the columns, between the columns and detector(s), and between the detector(s) and the waste reservoir.

General considerations for an HTSEC system are the same as for an SEC system used at room temperature except for the temperature control system. Special attention is needed to ensure baseline stability of the detector(s), to prevent erosion of equipment (especially columns) which come into contact with the high-temperature mobile phase, and to provide adequate ventilation for the mobile phase. Completely integrated, rather than modular systems are recommended for HTSEC.

Often SEC measurements are performed at column temperatures between 60–70 °C in order to decrease the viscosity of the mobile phase and to increase the solubility of the polymers. If the sample will dissolve in the mobile phase at room temperature, then an SEC system used at room temperature can be employed.

The Waters 150 C ALC/GPC instrument is microporcessor-controlled and operates well at temperatures up to 150 °C with harsh and very demanding solvents. It is designed for safety from solvent toxicity and flammability. The unit is entirely encased and has an efficient exhaust system. The instrument has three compartments: pump/solvent, injection and column/detector compartments. Each of which is independently thermostated. The differential refractometer is located in the column compartment and is thermostated. The other detectors are located externally and require heated extension tubing. Dissolution of polymer samples and injetion of the sample solution onto the column are carried out automatically. However, the Waters 150 C instrument is most reliable when used under constant experimental conditions; frequent changes of solvents and temperatures should be avoided.

The PL-GPC 210 is an automated system incorporating an injector, columns and a detector (a differential refractometer) in one oven and can be operated at temperatures up to 210 °C under microprocessor control. The solvent delivery module with a pump, a solvent reservoir and a degasser incorporated in a heated chamber. Sample solution can be injected automatically using an autosampler which is zone-heated to minimize oxidative degradation of samples. Similar to the Waters 150 C ALC/GPC, frequent changes of solvents and temperatures should be avoided. Operation with one solvent and at constant temperature is preferable.

3.3 Pumps and Injectors

3.3.1 Pumps

Solvent delivery pumps for HPLC have been developed to meet the requirements that the mobile phase can be delivered constantly at a flow rate of about 1 ml/min without pulsations under a pressure of 100 to 300 kg/cm^2. A constant, reproducible flow is very important in SEC. Most commercially available pumps for SEC are the positive displacement type which are similar to the so-called reciprocating pumps and are a kind of miniature syringe pump. A single-head positive displacement pump is shown schematically in Fig. 3.2. The plunger is actuated by a screw-feed drive through a gear box or by a sinusoidal cam, actuated by a stepping motor. The rate of solvent delivery by the plunger is controlled by the voltage on the motor.

The flow rate of the pump can be set digitally every 0.1 or 0.01 ml/min and the digital value is interlocked to the voltage on the motor, that is, the motor torque is proportional to the volume output of the pump. The movement of the plunger is linear at a specified rate with time during most of its travel in the pumping direction and results in a constant flow rate. The volume of the cylinder (the pumping chamber) is usually between 0.05 and 0.1 ml. After most of the solvent in the cylinder is delivered at the end of the stroke, the plunger moves back quickly (about 200 ms) to refill solvent. This process is repeated. When the effective volume of the cylinder is 0.05 ml, a flow rate of 1.0 ml/min can be attained by repeating this process 20 times in one minute. Solvent delivery is stopped in one movement at the refill process, causing pulsations.

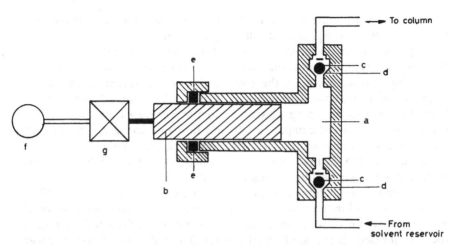

Fig. 3.2. Schematic diagram of a single-head positive displacement pump. *a* cylinder, *b* plunger, *c* check valve ball, *d* check valve seating, *e* plunger seal, *f* stepping motor, *g* gear box

In order to reduce the flow pulsations, dual-head or triple-head pumps having two or three cylinders are also available. Quick refill of solvent into the cylinders is not required in these types of pump: one cylinder fills solvent while the other(s) provides flow to the column. Sinusoidal cams or gears designed to produce a linear displacement of the plunger are used and the cycle of the solvent delivery is arranged to be slightly longer than the cycle of solvent refill to minimize the slight pulse at the end of the changeover points in the pumping cycle.

Rather inexpensive dual-head pumps having two different types of pumping heads are also commercially available: one (the first head) having two check valves at the inlet and outlet ends and the other (the second head) having no check valves. The volume of the cylinder of the first head is twice that of the second head. These two heads are connected in series and half of the solvent delivered from the first head passes to the column while the rest is taken up in the second head (accumulator head). After the delivery of the first head is complete, the plunger of the first head returns to the refill position. During this refill stroke, the second head discharges the remaining solvent to the column. This type of pump is called an accumulator pump.

In simple, single-head, reciprocating pumps, solvent delivery and refill processes are achieved by reciprocating the plunger at the same speed, actuated by a biased cam. The flow rate provided by a reciprocating pump can be changed by changing the plunger travel distance. As a biased cam is rotated 50 or 60 times per minute by a motor, depending on the cycle of electricity used, the plunger reciprocates 50 or 60 times per minute. Pulsations are greater with this type of pump and a pulse damper is generally used to minimize detector noise. Syringe pumps are similar to a large syringe and have a large cylinder volume of about 200 ml. Therefore, the volume is sufficient for one or several chromatograms without flow interruption (please note that some authors define this syringe pump as a positive displacement pump).

Output pressures up to 500 kg/cm² and maximum volumetric flow rate of 10 ml/min are typical of most commercially available pumps for HPLC. Because of this high-pressure output, the compression of the solvent occurs at the beginning of the pump stroke and simple reciprocating pumps deliver a slightly lower flow at high pumping pressures. Most sophisticated pumps can compensate for the solvent compressibility and flow rate changes. The flow rate from a syringe pump is more sensitive to solvent compressibility.

In order to keep the flow rate constant at high pressure, the roles of a check value and a plunger sealing are important. A check valve consists of a check valve ball made of sapphire or ceramic and a check valve seating made of PTFE. A check valve ball in the inlet check valve makes close contact with the check valve seating at the cycle of solvent delivery to prevent a solvent leaking back to the solvent reservoir and, similarly, the outlet check valve acts to prevent a back flow of solvent from the column system to the cylinder at the solvent refill cycle. A plunger seal made of solvent-resistant elastomer is placed between the plunger and the cylinder wall and prevents a solvent leaking from the cylinder to outside the head.

General pump specifications are: (1) pump resettability, (2) flow-rate accuracy, (3) flow-rate precision, (4) flow-rate stability, (5) pump pulsations, (6) output pressure, (7) maximum flow rate, and (8) chemical stability to solvents used as mobile phase. Pump resettability (or repeatability) means the ability to reset the pump to the same apparent flow rate repeatedly. Flow-rate accuracy relates to the ability of pumps to deliver exactly the same flow rate indicated by the pumping setting. Flow-rate precision is the same as flow-rate reproducibility and is a measure of flow rate consistency from one run to another. Flow-rate stability means the drift of the flow rate over relatively long periods.

When accurate determination of MW averages is desired, flow-rate precision and stability are the most important factors for SEC. Most modern pumps comply well with other specifications: pump resettability, small pump pulsations, high output pressure, flow rate up to 10 ml/min (usually 1 ml/min is sufficient), and compatibility with most organic solvents and aqueous solutions. The use of a stepping motor in pumps makes for easy pump resettability. Flow-rate accuracy is not so important if the specifications for flow-rate precision and stability are met. However, for semimicro or microbore SEC, short-term flow-rate precision (flow-rate reproducibility over several minutes) also becomes an important factor.

In order to maintain pumping systems, several precautions are required. The solvent used as the mobile phase should be filtered with a membrane filter of small pore such as 0.5 μm or an inlet filter should be placed at the pump inlet. The presence of small particles disturb the contact of the check-valve ball and the check-valve seating, and results in leakage or back flow of the mobile phase. The material for a pump seals should be resistant to the solvent used as the mobile phase. Elastomers which can be used for SEC are limited because common solvents used for SEC also dissolve most polymers. Wide-bore tubing is preferable at the pump inlet.

3.3.2 Injectors

The most generally useful sampling device for SEC is a six-port sample injection valve in which the sample is contained in an external loop. The volume of the loop is usually between 0.05 and 0.5 ml depending on the size of the columns used in the system and stainless-steel tubing of 0.25 or 0.5 mm i.d. is employed. The loop is filled with the sample solution by means of an ordinary syringe at the sample-load position. The excess volume of the sample solution should be flushed in order to thoroughly fill the loop. A clockwise rotation of the valve rotor to the sample-inject position diverts the mobile phase flow through the loop, and onto the column. Some injectors have a cavity on the disc instead of a loop. A constant injection volume is essential in SEC.

Automatic sample injectors, so-called autosamplers or autoinjectors, are commercially available so that large numbers of samples can be analyzed routinely without need for operator intervention. Care must be taken that the sample containers used in the autosampler are tolerant to the solvent and that the solvent does not evaporate while the sample solution is kept in the autosamplor before injection.

3.4 Detectors

3.4.1 Introduction

Detectors are used to monitor concentration changes in the column effluent by measuring the concentration or weight of solutes in the column effluent. For convenience, detectors are usually classified into three categories: concentration-based detectors, structure-selective detectors, and molecular-weight-sensitive detectors.

Concentration-based detectors are considered to be "universal" detectors which measure a change in a particular physical property of the column effluent, such as its refracture index. A viscometer also measures a change in a physical property of the column effluent but also responds to the MW as well as the concentration of the solute. Thus a viscometer is classified as a MW-sensitive detector. A light scattering detector is sensitive to both MW and the concentration of the solute in the effluent and also is classified as a MW-sensitive detector. Structure-selective detectors measure the properties specific to the solute and are sensitive only to sample components. Ultraviolet (UV) and infrared (IR) absorption photometers, fluorescence detectors, mass spectrometers, and nuclear magnetic resonance spectrometers are classified as structure-selective detectors.

Several requirements for the performance of SEC detectors are as follows: (1) The output signal that is provided in response to a change in solute concentration should be proportional to the amount of solute in the column effluent. Its detector response, except for MW-sensitive detectors, should be proportional to the amount of solute or the specific property of the solute and be independent of MW. (2) High signal-to-noise ratio, i.e. good baseline stability, is required. Detector drift should be minimized. (3) Wide dynamic range. The output signal should be linear over a wide concentration range. (4) The output signal in MW-sensitive detectors should be related to the MW of the solute. (5) Band broadening due to the detector should be minimized as much as possible. The volume of the detector cell should not be greater than one tenth the volume of the peak of a monodisperse solute.

3.4.2 Concentration-Based Detectors

(a) Differential Refractometer. The differential refractometer (a refractive index (RI) detector), detects the amount of solute in the column effluent by measuring the difference in the RI between the mobile phase and the column effluent containing the solute. This difference is proportional to the concentration of the solute. The measurable range of an RI detector is between 5×10^{-6} and 5×10^{-3} RIU full scale.

The refractive index of a sample solution (n) can be expressed as follows if it is a dilute solution

$$n = n_0 + (n' - n_0) c \tag{3.3}$$

where n_0 is the RI of the solvent, n' that of the sample and c is the concentration of the solution. As n_0 and $(n' - n_0)$ are constants in the sample solution, the RI of the sample solution becomes a function of the concentration. By rearrangement of Eq. (3.3) to

$$(n - n_0) = (n' - n_0) c \tag{3.4}$$

it becomes obvious that the difference in RI between the solvent and the sample solution is proportional to the sample concentration.

The RI detector is sensitive to all compounds in a solvent that have refractive indexes different from that of the solvent. Therefore, by choosing the solvent properly the RI detector is considered to be a universal detector.

Specific refractive index, that is the increment of refractive index to the concentration of a solute, dn/dc, is dependent on the MW of the solute and is expressed as

$$dn/dc = a + b/M_n \tag{3.5}$$

where a and b are constants and M_n is the number-average MW. For example, the differences in dn/dc between polystyrenes (RI about 1.60) of MWs 2000 and 1.8×10^6 in toluene (RI = 1.496) is 9%, whereas the difference in methyl ethyl ketone (RI = 1.381) is 4%. The larger the difference in RI between a solute and a solvent, the smaller the MW dependence of dn/dc. In general, when the degree of polymerization of a solute is greater than 10 monomer units or the MW is greater than 1000, the RI of the solute is practically independent of the MW of the solute and the response of the RI detector is a measure of the concentration or the weight of the solute in the effluent.

Of the three types of RI detector commercially available, the deflection type is the most commonly used in SEC. Fig. 3.3 illustrates its principle of operation. A light beam from a light source (a tungsten filament lamp) passes through an optical mask and is collimated by a lens. The parallel beam passes through a cell that consists of sample and reference sections separated by a dividing glass plate. The incident beam is deflected at the cell, reflected by the mirror, and again deflected as it passes back through the cell. The light beam then passes through the lens and on to the photocell via the optical zero glass. The photocell (a cadmium sulfide cell) is position-sensitive and the light beam strikes the surface of the photocell where it produces an electrical signal proportional to the position of the light beam on the photocell. Small changes in RI between the sample and reference sections of the cell cause a deflection in the location of the light beam on the surface of the photocell. The location of the beam, rather than its intensity, is determined. The size of the deflection is proportional to the concentration of solute in the column effluent.

The optical zero glass deflects the beam from side to side to adjust for zero output signal. The optical zero knob can turn about 20 times. The most sensitive position of the RI detector is about 10 turns from the maximum (the middle of the range of the optical zero adjustment) and the linearity of the detector is at this position. If the optical zero position is attained at the maxi-

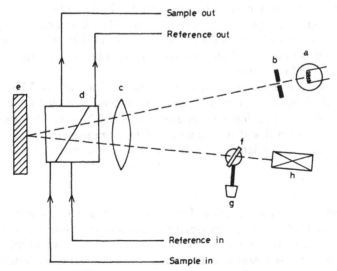

Fig. 3.3. Schematic diagram of a deflection-type differential refractometer. *a* light source, *b* mask, *c* lens, *d* cell, *e* mirror, *f* optical zero glass, *g* optical zero adjust knob, *h* photocell

mum of the knob turn when a solvent having lower or higher RI is used as the mobile phase, then the sensitivity decreases, the maximum case being one-tenth. When the position of the optical zero glass is inadequate to obtain the desired zero signal, mirror adjustments must be made by turning the mirror adjust screws slightly.

The second type of RI detector is based on Fresnel's law of reflection. This law states that the amount of incident light reflected at a glass–liquid interface is dependent on the angle of the light and the difference in refractive index between the glass and the liquid interfaces. In a Fresnel-type refractometer, the angle of incident light to the glass–liquid interface is near the critical angle where the Fresnel relationship becomes linear. A glass prism has two cavities to form two glass–liquid interfaces. One interface is used for the sample solution and the other for the reference liquid.

The third type of RI detector employs the shearing interferometer principle. An incident beam of known wavelength (e.g. 546 nm) is split into two beams of equal intensity by the beam splitter and these beams pass through the sample and reference cells, respectively, followed by re-focusing onto a second beam splitter. A difference in RI between sample and reference cells produces a difference in the optical path length which is measured by the interferometer.

The advantages of a deflection-type RI detector are: (a) a wide range of linearity, (2) only one prism is needed for the entire refractive index range, (3) relatively insensitive into contamination on the sample cell windows. Although a Fresnel RI detector cell can be made as small as 3 μl (the cell volume for a conventional deflection-type RI detector is about 10 μl), two different prisms must be used to cover the entire range of refractive index because of the restricted range of linearity. While a deflection-type RI detector measures the amount of deflection of the incident beam at the liquid cell, a Fresnel-type RI detector measures the intensity of the reflected light, so that the cell windows must remain clean for proper detector response. The interferometer-type RI detector is more sensitive than the other two types of detector, but the linear working range is rather small and several cells must be used in order to cover the entire range of refractive index.

RI detectors are sensitive to small temperature fluctuations. The thermal expansion coefficient of a liquid K_C is approximately 1×10^{-3} for common organic liquids such as chloroform and toluene and 0.21×10^{-3} for water. The refractive index of a liquid is a function of temperature and the change in refractive index Δn is related to K_C and to the change in temperature ΔT as follows:

$$-\Delta n = (n_0 - 1)K_C \Delta T \qquad (3.6)$$

For example, if the refractive index of an organic solvent is 1.4000 at 25 °C, then it increases to 1.4001 at 24 °C. Provided that the refractive index of a solute is 1.50, then the refractive index of a 0.1 % solution of the solute in the solvent becomes 1.4001. This means that measurement of the refractive index of the solution down to 1×10^{-6} RIU (1 % relative standard deviation) requires a temperature stability of 0.01 °C.

A temperature adjustment up to 0.01 °C is almost impossible for routine work at this time. Fortunately, an absolute measurement of the refractive index is not necessary and the differential measurement of refractive index between sample and reference solutions compensates for problems which may arise from temperature fluctuations. Moreover, in the case of MW calculations, the SEC chromatogram is normalized, therefore, the influence of temperature fluctuations are mostly compensated. Although most RI detectors have a temperature controller, it is recommended that the detector is located where it will not be subjected to drafts or excessive temperature variation. When water is used as the mobile phase, the effect of temperature on the refractive index is as low as one-tenth and the baseline drift is small as compared with organic solvents.

Baseline drift or a noisy baseline is caused by changes in solvent composition or density in the cell of the RI detector. In SEC, the most probable causes are: (1) a change in the concentration of air dissolved in the mobile phase (Air solubility in the solvent depends upon the temperature of the solvent.), (2) a change in temperature of the solvent in the reference section, (3) a bleed of materials from the column or from pump components, and (4) a change in temperature of the mobile phase. Degassing the mobile phase before use and keeping the temperature of the mobile phase constant are important ways to minimize baseline drift and a noisy baseline.

(b) Evaporative Light Scattering Detector. An evaporative light scattering detector is a type of universal detector, in which all the solvent from the column effluent is stripped by evaporation and the remaining particles are detected by light scattering. All materials less volatile than the solvent can be detected and the response of the detector is proportional to the mass of the sample in the solvent, thus it is also called a mass detector.

The effluent is atomized into a gas stream with a nebulizer and the atomized liquid is passed through a heated tube (an evaporator) where the solvent from the resultant droplets evaporates. All the solutes less volatile than the solvent are carried down the evaporator as an aerosol of fine particles. A light source and photomultiplier are arranged at the bottom of the evaporator, perpendicular to the flow, and the solute particles scatter the light from the light source. This scattered light is collected at a 60° forward angle by the photomultiplier. The integrated scattering intensity is related to the concentration of the solute(s) in the effluent.

There are three models commercially available. A halogen lamp or a He/Ne laser is used as the light source and a photomultiplier or a silicon photodiode is used to detect the scattered light. Careful selection of both the temperature of the evaporator and the flow rate of the nebulizing gas is required to optimize the response of the detector. The temperature of the evaporator is dependent on the type of sample and the mobile phase: normally 40–50 °C may be sufficient. One manufacturer recommends the following temperatures: methylene chloride 70 °C, tetrahydrofuran 100 °C, water 150 °C. Nitrogen or compressed air is used as the nebulizing gas at about 20 psi or 5 l/min, for example. For safety considerations, nitrogen should be used.

(c) **Flame Ionization Detector.** The other type of detector based on the concept of removing the mobile phase prior to detection is a flame ionization detector. A fine stream of effluent from the column is deposited on a moving carrier (stainless-steel wire, quartz fiber belt, etc.). The solvent is then evaporated in a heated chamber (an evaporative oven). The moving carrier coated with the solute passes into a combusion unit (a pyrolyzer oven) where gaseous products from the solute are purged with nitrogen gas into a flame ionization unit. In the alternative combustion unit, the moving carrier is passed directly through a H_2/air flame to generate a signal. The moving carrier is cleaned by passing it through a cleaning oven for the next effluent deposit.

(d) **Density Detector.** A density detector measures the increase in density in the column effluent. The mechanical oscillator method is applied and an oscillating, U-shaped glass tube is used as the measuring cell. The density d of a sample solution in the oscillating tube is related to the square of the period of oscillation T of the filled cell by the following equation:

$$T^2 = A + Kd \qquad (3.7)$$

where A and K are constants for the individual cell at a given temperature and can be determined from frequency measurements with a fluid of known densities.

The difference in densities between a polymer solution d_2 and pure solvent d_1 can be written as

$$d_2 - d_1 = c(1 - V_2^* d_1) \qquad (3.8)$$

where c is the polymer concentration and V_2^* is the apparent specific volume of the polymer solute. The difference in densities is also given by

$$d_2 - d_1 = (1/A)(T_2^2 - T_1^2) \qquad (3.9)$$

where T_1 and T_2 are the periods of oscillation for the polymer solution and the pure solvent, respectively. Consequently, in SEC analysis, the concentration of the solution eluting at a given retention volume is

$$c = (T_2^2 - T_1^2)/A(1 - V_2^* d_1) \qquad (3.10)$$

When the differences between d_1 and d_2 and T_1 and T_2 are small, $(d_2 - d_1)$ is proportional to $(T_2 - T_1)$. Thus, a plot of $(T_2 - T_1)$ as a function of retention volume provides a similar chromatogram for the polymer as does the RI chromatogram.

A density detector is similar to an RI detector in nature and the density of the solution (the column effluent) depends on temperature and MW of the solute. A density detector can be applied to the analysis of olefinic polymers and other non-UV-absorptive samples as an alternative to an RI detector. The most typical application is the analysis of chemical heterogeneity of copolymers which consist of both UV- and non-UV-active comonomers.

3.4.3 Structure-Selective Detectors

(a) **UV Detector.** Bonding electrons absorb radiation of specific wavelengths in the 190–800 nm range. Ultraviolet (UV) and visible (VIS) absorption detectors measure the absorption of radiation in the wavelength range between 190 and 350 nm for UV and 350 and 800 nm for VIS according to the Lambert–Beer law, defined as follows:

$$\text{Absorbance} = \log(I_0/I) = \varepsilon b c \qquad (3.11)$$

where I_0 is the monochromatic incident light beam, I is the transmitted beam, ε is the molar absorption coefficient, b is the pathlength (cm) of the cell, and c is the solute concentration (M).

The simplest UV detector is a single-wavelength UV detector. This uses a low-pressure mercury lamp as a light source which emits radiation at 254 nm which is easily isolated using simple optical filters. This detector can only be used at a wavelength of 254 nm. Detection at a wavelength of 280 nm or other wavelengths is also possible by changing the optical filters. The most commonly available UV detector is a variable-wavelength UV detector which utilizes a deuterium lamp as the light source, coupled with a diffraction grating mono-chromator which selects the desired wavelength. A tungsten lamp is used for visible light. Typical dimensions of the flow cell are a pathlength of 10 mm and a bore of 1 mm, resulting in a volume of 8 μl.

A photodiode array UV-VIS detector employs a photodiode array to measure the intensity of the transmitted light. The photodiode array consists of a corresponding number of photoreceptors mounted on a 1-cm silicon chip and detects the intensity of transmitted light at all wavelengths simultaneously, each photoreceptor receiving a different wavelength. An array of 512 photo-diodes, for example, detects the transmitted radiation over a 455 nm range (195–650 nm) at an interval of about 1 nm. Usually, the response from several photodiodes are accumulated, so that the spectral band width is more than 1 nm (e.g. 2, 4, 8, or 16 nm).

UV-VIS detectors are solute specific and are used only if the sample has chromophores which absorb at the particular wavelength of the detector. The mobile phase should be transparent at the wavelength of interest. The molar absorption coefficient is the inherent value of the material at the specified wavelength. Compounds having a phenyl group or a double bond have values of ε between 100 and 10^4. In general, a UV detector is more sensitive than a RI detector. However, in the case of polymers (e.g. polystyrene), ε is not very high (about 220 for polystyrene at 254 nm) and the sensitivity of the UV detector is comparable to that of an RI detector. Highly sensitive detection and a low detection limit are attained if the compound has a higher value of ε. The value of ε is dependent on MW, but those of polymers and oligomers having MWs over 1000 are considered to be independent of MW as in the case of an RI detector.

With polystyrene a UV wavelength of 254 nm can be used with tetra-hydrofuran or chloroform as the mobile phase. Toluene and methyl ethyl ketone cannot be used as the mobile phase because of their UV absorption. When poly(methyl methacrylate) or other methacrylate and acrylate polymers are analytes, a UV detector can be used at wavelengths around 233 nm with 1,2-dichloroethane as the mobile phase.

(b) **IR Detector.** The vibrations of molecular bonds absorb radiation of specific wavelengths in the infrared region. Two types of infrared (IR) detector are available: dispersion and non-dispersion types. The dispersion-type IR detector uses filters or a monochromator between the sample cell and the detector of the IR radiation. Single- and double-beam IR detectors operating over the wavelength range $2.5-14.5$ μm ($4000-690$ cm^{-1}) or $2.5-25$ μm ($4000-400$ cm^{-1}) are available commercially.

The values of ε in IR absorption are small and are in general less than 100. Exceptions are carbonyl groups (about 600) and a several other groups which are not common in polymers and oligomers. Satisfactory sensitivity can be obtained by monitoring the C–H streching frequency around $2850-2950$ cm^{-1} and the C=O frequency around $1650-1750$ cm^{-1}. The mobile phase should be a chlorinated organic solvent that has suitable spectral windows in these regions.

A Fourier-transform infrared (FTIR) spectrophotometer is used as a non-dispersion-type IR detector. In general, a flow-through cell is used to monitor the effluent continuously as in the case of dispersion-type IR detectors. The major limitation of IR detection in SEC is that the most common solvents used as the mobile phase also have IR absorption at several wavelengths and, therefore, it is difficult to select a solvent with suitable spectral windows in the IR region.

Recently interfaces that remove the solvent from the effluent to detect the solute directly by FTIR have been developed. An example of such an interface is when the effluent from the column is sprayed with a heated nitrogen stream through an ultrasonic nozzle to a Ge disc which is placed in a vacuum chamber. The solute in the effluent deposits on the Ge disc and the IR spectrum of the solute on the Ge disc is measured. Polymer samples form films on the Ge disc. However the morphology of the deposited polymer film may affect the IR spectra.

3.4.4 MW-Sensitive Detectors

Light scattering detectors and viscosity detectors are dealt with in detail in Chap. 8.

4 Columns

4.1 Introduction

SEC column packing materials should be rigid and porous and have pore sizes ranging from approximately 30 to 4000 Å. There are three different types of packing materials: semirigid polymer gels, rigid inorganic solids, and soft polymer gels. The required performance of the packing materials involves mechanical, chemical, and thermal stability, high-resolution separation, and low back pressure when packed in columns. Chemical inertness of the packing materials, not only to the mobile phase but also to solutes, is one of the important factors for SEC. In order to maintain high-resolution separation, some of these packing materials are available only in the form of packed columns.

High resolution and low back pressure in a packed column are in antinomy with each other. In order to obtain high-resolution separation, in other words, to obtain a highly efficient column, the particle size of packings should be low and the particle size distribution should be as narrow as possible. Modern SEC uses packing materials of small particle sizes, which results in an increase in back pressure in a packed column. Therefore, the rigidity of the packing materials is also one of the important factors for modern SEC. When soft organic gels are used as packing materials, a solvent delivery pump cannot be used and a Mariotte flask or equivalent has to be employed to regulate the flow of the mobile phase. SEC with soft organic gels is often referred to as gel chromatography.

Solute molecules in SEC should be separated soley by size exclusion without any secondary interactions with gels. This means that the separation of solutes is regulated by the selection of the pore size of the packing material, not by the mobile phase composition, and adequate selection of the pore size of the packing material is essential for SEC.

4.2 Types of Packing Materials

Spherical semirigid, highly cross-linked polystyrene (PS) gels (styrene–divinyl-benzene copolymers) are widely used for SEC analyses of synthetic polymers and oligomers which are soluble in organic solvents. Water-soluble polymers can be analyzed by aqueous SEC and a variety of hydrophilic semirigid polymer gels for modern SEC are available in the form of packed columns: poly(glyceryl methacrylate) gel, poly(2-hydroxy methacrylate) gel, poly(vinyl alcohol) gel, poly(styrene sulfonate) gel and polyacrylamide gel.

Silica-based packing materials are rigid and have several advantages over the semirigid polymer gels: e.g. mechanical stability and a much wider range of usable mobile phases. However adsorption of solutes onto the surface of inorganic materials and the limitation of the available porosity of the packing materials are potential disadvantages. Because of the activity of silanol groups on the surface of silica gel, chemical derivatization on the surface is required to reduce the adsorption of solutes on the surface. For example 1,2-epoxy-3-propoxypropyltriethoxysilane is chemically bonded onto the silica surface, followed by opening the epoxide ring to form dihydroxy groups, so that the silica surface is covered with the hydrophilic diol functional groups. Chemical modification of the silica surface with chlorotrimethylsilane, for example, creates a nonpolar, hydrophobic surface which can be used for non-aqueous SEC.

Table 4.1. Properties of soft polymer gels available in bulk

Designation[a]	Fractionation range (dextran)	Designation	Fractionation range (globular protein)
Sephadex		Biogel	
G-10	< 700	P2	100–1800
G-15	< 1500	P4	800–4000
G-25	100–5000	P6	1000–6000
G-50	500–10000	P10	1500–20000
G-75	1000–50000	P30	2500–40000
G-100	1000–100000	P60	3000–60000
G-150	1000–150000	P100	5000–100000
G-200	1000–200000	P200	30000–200000
Sepharose		P300	60000–400000
2B	$1 \times 10^5 - 2 \times 10^7$	Biogel A	
4B	$3 \times 10^4 - 5 \times 10^6$	0.5 m	$< 10000 - 5 \times 10^5$
6B	$1 \times 10^4 - 1 \times 10^6$	1.5 m	$< 10000 - 1.5 \times 10^6$
		5 m	$10000 - 5 \times 10^6$
Sephacryl		15 m	$40000 - 1.5 \times 10^7$
S-100HR	not available	50 m	$100000 - 5 \times 10^6$
S-200HR	$1 \times 10^3 - 8 \times 10^4$	150 m	$1 \times 10^6 - 1.5 \times 10^8$
S-300HR	$2 \times 10^3 - 4 \times 10^5$		
S-400HR	$1 \times 10^4 - 2 \times 10^6$		
S-500HR	$4 \times 10^4 - 2 \times 10^7$		
Fractogel			
HW-40	100–7000		
HW-50	500–20000		
HW-55	1000–200000		
HW-65	$10000 - 1 \times 10^6$		
HW-75	$1 \times 10^5 - 1 \times 10^7$		

[a] Supplier and particle sizes: Sephadex – Pharmacia, 20–50 or 40–120 µm; Sepharose – Pharmacia, 45–165 µm; Sephacryl – Pharmacia, 25–75 µm; Biogel P – Bio-Rad, 40–80 µm; Biogel A – Bio-Rad; Fractogel – Toso, type F 32–63 µm, type S 25–40 µm.

Soft polymer gels are not utilized in modern SEC in the form of packed columns because these gels collapse at high inlet pressure. Sephadex – cross-linked dextrans, Sepharose – agarose matrix, Sephacryl – cross-linked copolymer of allyldextran and N,N'-methylene bisacrylamide, Biogel P – copolymer of polyacrylamide and N,N'-methylene bisacrylamide, Biogel A – agarose gel, and Toyopearl (Fractogel) – polyether gel, are available. These soft gels are mostly hydrophilic and are used in aqueous media. Some of them can be used with polar organic mobile phases. They are sold in a bulk form as powder or in a swollen state and can be packed in a glass column by the user. Table 4.1 lists examples of these soft gels.

4.3 Non-aqueous SEC Columns

Columns packed with spherical porous PS gels of 5 – 7 µm diameter are available in a variety of pore sizes from several suppliers. Table 4.2 lists the characteristics of these columns. The molecular weight at the exclusion limit is determined with polystyrene. The number of theoretical plates per column is between 12000 and 16000, depending on the particle diameter and the pore size of the PS gel. In general, these columns are packed in tetrahydrofuran or toluene. Besides the columns listed in Table 4.2, 50 or 60 cm columns filled with toluene, chloroform, or other organic solvents packed with PS gels of about 10 µm diameter are also available from the same suppliers. Wide-bore columns (e.g. 20 mm i.d.) are also available for preparative work. Particle diameters of PS gels packed in the preparative columns are larger than 10 µm and, therefore, the number of theoretical plates is less than those for analytical-scale columns.

Three different pore sizes of porous silica microsphere (particle diameter 6 µm) chemically bonded with trimethyl groups are available in the form of packed columns from Hewlett-Packard: PSM-60 (pore size 60 Å), PSM-300 (300 Å), and PSM-1000 (1000 Å). Column dimensions are 6.2 mm i.d. and 25 cm length. Porous spherical silica microparticles (particle diameter 10 µm) are also available in the form of packed columns from Merck: LiChrosphere Si100 (pore size 100 Å), Si300 (300 Å), Si500 (500 Å), Si1000 (1000 Å), and Si4000 (4000 Å). Column dimensions are 4 mm i.d. and 25 cm length.

The pore size of PS gels cannot be measured by mercury porosimetry as in the case of rigid silica gels. Instead, the MW of PS which appears at the exclusion limit is employed to specify porosity of the PS gels. PLgel and Ultrastyragel columns are designated with nominal porosity in Å units. However, these values are *not* the real pore sizes of the PS gels, but are the extended chain lengths of PS which appear at the exclusion limit. The product of the nominal porosity times 41.3 is approximately equal to the MW of PS at the exclusion limit. For example, PS gels packed in PLgel 100 Å do not have a pore size of 100 Å, but the MW at the exclusion limit of the gels is 4000 (100 Å times 41.3).

Small-pore gel columns are used to separate small molecules and oligomers. Large-pore gel columns can be used for the separation of polymers and two to four columns of different pore sizes typically are connected in series. Mixed-

Table 4.2. Non-aqueous SEC columns packed with PS gels

Designation[a]	MW exclusion limit (PS)	Designation[a]	MW exclusion limit (PS)
Shodex		Ultrastyragel (cont.)	
KF-801	1.5×10^3	10^6A	(1×10^7)
KF-802	5×10^3	Linear	4×10^6
KF-802.5	2×10^4	PLUS 10^3A	3×10^4
KF-803	7×10^4	PLUS 10^4A	6×10^5
KF-804	4×10^5	PLUS 10^5A	4×10^6
KF-805	4×10^6	PLUS 10^6A	(1×10^7)
KF-806	(2×10^7)		
KF-807	(2×10^8)	TSKgel	
KF-806M	(2×10^7)	G1000H$_{XL}$	1×10^3
KF-803L	7×10^4	G2000H$_{XL}$	1×10^4
KF-804L	4×10^5	G2500H$_{XL}$	2×10^4
KF-805L	4×10^6	G3000H$_{XL}$	6×10^4
KF-806L	(2×10^7)	G4000H$_{XL}$	4×10^5
KF-807L	(2×10^8)	G5000H$_{XL}$	4×10^6
		G6000H$_{XL}$	(4×10^7)
Gelpak		G7000H$_{XL}$	(4×10^8)
GL-A110	1×10^3	GMH$_{XL}$	(4×10^8)
GL-A120	5×10^3		
GL-A130	2×10^4	PLgel	
GL-A140	7×10^4	50A	2×10^3
GL-A150	5×10^5	100A	4×10^3
GL-A160	4×10^6	500A	2×10^4
GL-A170	(5×10^7)	10^3A	4×10^4
GL-A100M	(5×10^7)	10^4A	4×10^5
		10^5A	4×10^6
Ultrastyragel		10^6A	(2×10^7)
100A	1.5×10^3	MIXED-A	(4×10^7)
500A	1×10^4	MIXED-B	(1×10^7)
10^3A	3×10^4	MIXED-C	2×10^6
10^4A	6×10^5	MIXED-D	4×10^5
10^5A	4×10^6	MIXED-E	3×10^4

[a] Column size [i.d. (mm) × length (cm)] and supplier: Shodex – 8 × 30, Showa Denko; TSKgel – 7.8 × 30, Toso; Gelpak – 10 × 30, Hitachi Chemical; PLgel – 7.5 × 30, Polymer Laboratories; Ultrastyragel – 7.8 × 30, Waters. The figures in parentheses represent approximate values.

pore gel columns in which gels of several different pore sizes are mixed are also available: Shodex KF806M, TSKgel GMH$_{XL}$, Gelpak GL-100M, PLgel MIXED, and Ultrastyragel Linear. The mixing ratio of gels of different pore sizes are adjusted so as to make approximately linear calibration plots. Examples of PS calibration curves for Shodex KF columns are shown in Fig. 4.1.

Mixed-pore gel columns have approximately linear calibration plots, but the plots are still slightly curved. Similarly, the top and the bottom of the calibration curves for most single-pore gel columns are curved upwards and downwards, respectively. Columns having literally linear calibration plots are commercially available: Shodex KF-803L to KF-807L, PLgel MIXED-A to MIXED-E, and

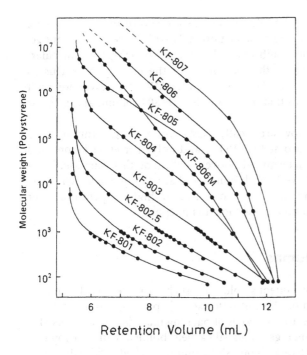

Fig. 4.1. MW calibration curves for Shodex KF columns. Mobile phase: tetrahydrofuran; sample: polystyrene. (Reprinted by courtesy of Showa Denko)

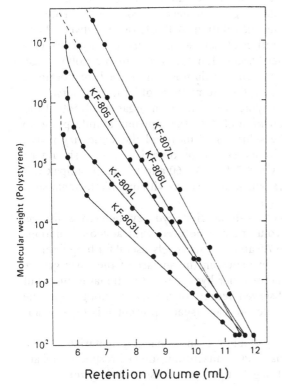

Fig. 4.2. MW calibration curves for Shodex KF SEC linear series columns. Mobile phase: tetrahydrofuran; sample: polystyrene. (Reprinted by courtesy of Showa Denko)

Ultrastyragel PLUS10³A to PLUS10⁶A. Examples are shown in Fig. 4.2. These columns are prepared by mixing smaller-pore gels with these single-pore gel columns in order to make their calibration plots linear up to the low-molecular-weight region without changing the magnitude of the MW at the exclusion limit. For example, the Shodex KF-804L column has a MW at the exclusion limit of 4×10^5 which is the same as that of the Shodex KF-804 column and is linear down to a MW of 100.

The columns discussed above are mostly used at room temperature or up to 60 °C. Polyolefins are measured at 130–150 °C and polyether ether ketone and polyphenylene sulfide are determined above 200 °C. For this purpose, high-temperature SEC columns have been designed and are commercially available. The abbreviations HT (high temperature) or UT (ultra-high temperature) after the designation of each column are used for this purpose.

4.4 Aqueous SEC Columns

Aqueous SEC uses water or aqueous solutions such as buffes, as the mobile phase for the determination of MWs of water-soluble polymers, thus packings should be hydrophilic. There are two types of packing material for aqueous SEC columns: hydrophilic polymer gels and silica gels bonded with hydrophilic functional groups. Table 4.3 lists aqueous SEC columns packed with hydrophilic polymer gels. These columns have approximately 10000–16000 theoretical plates. They can be used with aqueous solutions with pH values between 2 and 10 and are useful for the separation of natural and synthetic water-soluble polymers. The packing material Shodex Ionpak is an anionic polymer gel and Shodex CHITOpak is cationic. These columns can be used for non-ionic water-soluble polymers or can be used for some types of ionic polymers.

Although these hydrophilic polymer gels shrink in organic solvents, 10 to 20 % alcohol or acetonitrile can be added to the aqueous mobile phase to prevent adsorption between the solute molecules and the stationary phase. Excess addition of organic solvents or use of organic solvents should be avoided. Exceptions are Gelpak GL-W520 to W-560 and Shodex Asahipak GF-310HQ to GF-7M HQ in Table 4.3, which can be used with organic solvents, such as chloroform and tetrahydrofuran.

Examples of columns packed with organic hydroxyl-bonded silica gel are listed in Table 4.4. These columns can be used with aqueous solutions having pH values between 3 and 7.5 and are exclusively useful for biopolymers, such as proteins. Most of the active sites on the surface of the silica gel are blocked with glyceropropylsilane or other hydrophilic functional groups, but adsorptive effects usually are observed because of residual silanol groups. The addition of a small amount of alcohol can usually prevent this type of adsorption.

Besides the packed columns in Table 4.4, some packing materials can be obtained in bulk form. Synchropak GPC, Aquachrom, and Lichrosphere Diol are glyceropropyl-group-bonded silica gel. Superose is cross-linked agarose.

Table 4.3. Aqueous SEC columns packed with hydrophilic polymer gels

Designation[a]	MW exclusion limit (pullulan)	Designation[a]	MW exclusion limit (dextran)
Shodex OHpak		TSKgel	
SB-802HQ	4×10^3	G2500PW$_{XL}$	5×10^3
SB-802.5HQ	1×10^4	G3000PW$_{XL}$	2×10^5
SB-803HQ	1×10^5	G4000PW$_{XL}$	1×10^6
SB-804HQ	2×10^6	G5000PW$_{XL}$	2.5×10^6
SB-805HQ	4×10^6	G6000PW$_{XL}$	(5×10^7)
SB-806HQ	(2×10^7)	GMPW$_{XL}$	(5×10^7)
SB-806MHQ	(2×10^7)		
Q-801	1.8×10^3	PL aquagel	
Q-802	5×10^3	P2	1×10^4
		P3	1×10^5
ShodexCHITOpak			
KQ-802.5	7×10^3	Designation[a]	MW exclusion limit [poly(ethylene oxide)]
Shodex Ionpak			
KS-801	1×10^3		
KS-802	1×10^4	PL aquagel-OH	
KS-803	5×10^4	40	2×10^5
KS-804	4×10^5	50	1×10^6
KS-805	5×10^6	60	(2×10^7)
KS-806	(5×10^7)		
		Ultrahydrogel	
Shodex Asahipak		120	5×10^3
GS-220HQ	3×10^3	250	8×10^4
GS-320HQ	4×10^4	500	4×10^5
GS-520HQ	3×10^5	1000	1×10^6
GS-620HQ	2×10^6	2000	(7×10^6)
GF-310HQ	4×10^4	Linear	(7×10^6)
GF-510HQ	3×10^5		
GF-710HQ	1×10^7		
GF-7MHQ	1×10^7		
Gelpak			
GL-W510	2×10^3		
GL-W520	6×10^3		
GL-W530	5×10^4		
GL-W540	4×10^5		
GL-W550	2×10^6		
GL-W560	(5×10^7)		

[a] Column size [i.d. (mm) × length (cm)], packing material, and supplier: Shodex OHpak SB series – 8 × 30, poly(glyceropropyl methacrylate) gel, Showa Denko; Shodex OHpak Q series – 8 × 30, poly(vinyl alcohol) gel, Showa Denko; Shodex CHITOpak – 8 × 30, Chitosan gel, Showa Denko; Shodex Ionpak – 8 × 30, sulfonated PS gel, Showa Denko; Shodex Asahipak – 7.6 × 30, poly(vinyl alcohol) gel, Showa Denko; Gelpak – 10.7 × 30, hydrophilic polyacrylate gel, Hitachi Chemical; TSKgel – 7.8 × 30, hydrophilic polymer gel, Toso; PL aquagel – 7.5 × 30, polyacrylamide gel, Polymer Laboratories; PL aquagel-OH – 7.5 × 30, hydroxyl macroporous polymer gel, Polymer Laboratories; Ultrahydrogel – 7.8 × 30, hydroxy polymethacrylate gel, Waters. The figures in parentheses represent approximate values.

Table 4.4. Aqueous SEC columns packed with silica gel bonded with organic hydroxyl groups

Designation[a]	MW exclusion limit (protein)	Designation[a]	MW exclusion limit (protein)
Shodex PROTEIN		TSKgel	
KW-802.5	1.5×10^5	G2000SW$_{XL}$	1×10^5
KW-803	7×10^5	G3000SW$_{XL}$	5×10^5
KW-804	2×10^6	G4000SW$_{XL}$	(7×10^6)
μ-Bondagel			
E-125 A	5×10^5		
E-500 A	5×10^5		
E-1000 A	2×10^6		
Linear	2×10^6		
High A	7×10^6		

[a] Column size [i.d. (mm) × length (cm)], packing material, and supplier: Shodex – 8 × 30, glyceropropyl bonded silica gel, Showa Denko; TSKgel – 7.8 × 30, hydrophilic group bonded silica gel, Toso; μ-Bondagel – 7.8 × 30, ether-group bonded silica gel, Waters. The figures in parentheses represent approximate values.

4.5 Care and Handling

Separability of sample molecules in SEC is primarily dependent on the pore size of the packing material and is independent of mobile phase composition. This is the most important difference between SEC and interactive forms of HPLC. For example, molecules with a larger diameter than the packing material elute at the retention volume corresponding to the exclusion limit. These molecules cannot be separated. If a column packed with a gel of large pore size is used for the separation of oligomers with a MW of around 500, no separation occurs even though the column may have a high number of theoretical plates. A column packed with gel of a small pore size must be used in this case.

The peak width of an SEC chromatogram is a function of both the MWD of the sample and the peak broadening effects during the elution of the sample. Peak broadening effects and tailing due to the use of inefficient columns lead to erroneous results (see Chap. 2). The use of highly efficient columns, that is, columns having a high number of theoretical plates, is essential for SEC. To ensure maximum efficiency, care must be taken to keep the SEC columns in good condition.

Particular attention should be paid to the use of PS gel columns by using the following measures:

(1) Non-solvents for PS (not PS gel), such as water, alcohols, and hydrocarbons, cannot be used as mobile phases. A small amount of shrinkage of the PS gel packing material in a column will cause an irreversible void in the inlet of the column and result in a decrease in column efficiency. Restoration to the original condition is impossible. The use of mobile phases containing a small

amount of these solvents should also be avoided. For example, if 1% water is included in a THF mobile phase, then the elution of PS samples is retarded and that of polyethylene glycols is increased.

(2) When the solvent in a column has to be replaced with a different solvent, the flow rate should be decreased to less than 0.5 ml/min. Exchangeable solvents should be used that are good solvents for PS (not gel), e.g. tetrahydrofuran, chloroform, and toluene. These good solvents can be used alternately. However, it is recommended that the mobile phase is not replaced too often in order to keep the column highly efficient. Replacement of the filling solvent with dimethylformamide or hexafluoro-2-propanol is possible, but the reverse replacement should be avoided.

(3) Maximum flow rate and column pressure are typically 2.0 ml/min and 35 kg/cm^2, respectively. Excess shock to a column should be avoided. When the column temperature is raised, the pump should be stopped after lowering the column temperature to room temperature.

(4) Viscous sample solutions should not be injected into a column. Maximum sample concentrations are as follows depending on the MW of the polymers: MW below 5000, 1.0%; MW 5000 – 25000, 0.5%; MW 25000 – 2 × 10^5, 0.25%; MW 2 × 10^5 – 2 × 10^6, 0.1%; MW above 2 × 10^6, 0.05%.

(5) Storing columns at ambient temperature is recommended. Column-end fittings should not be disconnected in order to prevent the loss of packing material in the column. It is important to prevent the columns from drying out during storage and both ends of a column should be plugged tightly. It is recommended that solvents are replaced in columns periodically, or to immerse one end of the column into the solvent used as the mobile phase after deplugging one end. If tetrahydrofuran is used as the filling solvent for storage, a stabilizer, such as 2,6-di-*tert*-butyl-*p*-cresol (BHT), should be added. Replacement of tetrahydrofuran with toluene is recommended when the column is not used for a long period.

(6) The use of a precolumn is recommended to prevent irreversible adsorption of impurities from the sample polymers. Sample solutions and solvents used as mobile phases should be filtered using a microfilter of approx. 0.5 μm.

The same care and attention should be given to aqueous SEC columns. For columns packed with silica-based materials, the maximum flow rate and the column pressure are typically 1.5 ml/min and 50 kg/cm^2. The pH range of the mobile phase should be 3 to 7.5 and the column temperature less than 50 °C. For columns packed with polymer-based materials, the flow rate and column pressure should be lower than the values above. When aqueous SEC columns are stored for long periods, the solvent in the columns should be replaced with 0.05% aqueous sodium azide solution or 5 – 10% aqueous methanol solutions to prevent the growth of microorganisms.

5 SEC Method Development

5.1 Sample Preparation

5.1.1 Polymer Solubility

Polymer samples must be dissolved in the solvent used as the mobile phase. In non-aqueous SEC, solvents that can be used as the mobile phase are limited to those compatible with the stationary phase as well as those that are solvents for the polymer samples. However in special cases, samples can be dissolved in a different solvent from the mobile phase. For example, urea-formaldehyde resin dissolves in both N,N-dimethylformamide (DMF) and dimethylsulfoxide (DMSO), but DMSO is a much better solvent for the resin than DMF. Therefore, the resin is dissolved in DMSO and the solution injected into a system of polystyrene (PS) gel-stationary phase/DMF mobile phase. The other example is the SEC of isotactic polypropylene (iPP) with cyclohexane as the mobile phase at 70 °C. The sample iPP is insoluble in cyclohexane and is therefore dissolved in decalin at 140 °C and then diluted with hot cyclohexane and kept at about 70 °C prior to injection into the SEC system.

A "good" solvent is a solvent that can dissolve a polymer sample in any proportion at most temperatures. Also, it has a theta temperature (Flory temperature) that is well below room temperature. (The "theta temperature" is defined as the temperature at which the second virial coefficient of osmotic pressure of the polymer solution becomes zero and is the critical miscibility temperature at infinite molecular weight (MW) where the polymer and the solvent are no longer miscible.) A "poor" solvent is used to denote one in which the solubility of the polymer is limited. A poor solvent is also a solvent whose theta temperature is close to room temperature. A non-solvent is defined as a solvent whose ability to dissolve the polymer of interest is almost negligible.

Dissolving a polymer is a slow process that occurs in two stages. First, solvent molecules slowly diffuse into the polymer to produce a swollen gel. The rate-controlling step is the solvent diffusion rate within the swollen polymer. Warming the polymer/solvent mixture reduces solvent viscosity and speeds up diffusion. The second stage is the gradual disintegration of the gel into a true solution. Mechanical agitation can speed up this stage. The crystallinity of the polymer, the viscosity of the system, the size difference between polymer and solvent molecules, and the effects of the morphology and MW of the polymer make the solubility relations in a polymer system more complex than those for low-MW compounds.

A cross-linked polymer does not dissolve in any solvent and can only swell if it interacts with a solvent that can dissolve the non-cross-linked polymer. However, the absence of solubility does not imply cross-linking. High inter-molecular forces, such as crystallinity and hydrogen bonding, can prevent solubility. Many crystalline polymers dissolve in specific solvents at temperatures near their melting points. Linear polyethylene with a crystalline melting point of 135 °C is soluble in many solvents at temperatures above 100 °C. Polar crystalline polymers, such as polyamides, are soluble in solvents that interact strongly by hydrogen bonding with the polymers at room temperature (e.g. acidic solvents such as *m*-cresol and chlorophenol).

The solubility of a polymer sample in a solvent can be calculated from the solubility parameters of both the polymer and the solvent. The solubility parameter is defined by

$$\delta = (\Delta E/V)^{1/2} \tag{5.1}$$

where $\Delta E/V$ is the energy of vaporization per unit volume, or the cohesive energy density. This parameter can provide the measure of polymer–solvent interactions. If the solubility parameter of the solvent is equal to that of the polymer, then the two substances are mutually soluble. The tendency towards dissolution decreases with an increase in the difference between the solubility parameters of the solvent and the polymer.

The solubility parameter, however, is not a universal parameter for the calculation of polymer solubility. The solubility parameter of PS is 18.5 ($J^{1/2}/cm^{3/2}$) (mean value) and those of tetrahydrofuran (THF), chloroform, and toluene are 19.5, 19.0, and 18.2, respectively. Therefore PS is soluble in these solvents. Poly(vinyl chloride) has a solubility parameter of 19.7, but is not soluble in chloroform and toluene. Polyacrylonitrile has a solubility parameter of 25.7 and is soluble in DMF (24.8), but not in acetonitrile (24.3). Poly(ethylene terephthalate), which has a solubility parameter of 20.5, is not soluble in dioxane (20.5), but is soluble in hexafluoroisopropanol (less than 16.0). The solubility parameters of several common polymers and solvents are listed in Tables 5.1 and 5.2, respectively.

The solubility parameter approach is not a general procedure to be used to select solvents for polymers. Hydrogen bonding, the dielectric constant, and other related quantities must be taken into account. Sometimes it is advantageous to classify the ability of solvents to dissolve solutes by separating the contributions of three different solubility parameters: hydrogen bonding, dispersion, and polar forces. Poly(vinyl chloride) may require solvents which have higher polar force solubility parameters such as THF and cyclohexanol. Hexafluoroisopropanol is a weakly acidic solvent and can dissolve poly(ethylene terephthalate), even though the solubility parameter of this solvent is much lower than the polymer.

Biopolymer solubility is significantly altered by adjusting the salt concentration and pH levels of aqueous solvents. The ionic strength of aqueous

Table 5.1. Solubility parameters for some common polymers

Polymer	Solubility parameter	Polymer	Solubility parameter
Polyisobutylene	16.0 – 16.5	Poly(ethyl acrylate)	19.2
Polyethylene	16.2 – 17.1	Poly(vinyl acetate)	19.1 – 19.6
Polyisoprene	16.6 – 17.1	Poly(vinyl chloride)	19.3 – 20.1
Polybutadiene	16.6 – 17.6	Poly(ethylene terephthalate)	20.5
Poly(ethyl methacrylate)	18.3 – 18.6	Poly(methyl acrylate)	20.7
Polypropylene	18.8 – 19.2	Polyacrylonitrile	25.3 – 26.1
Polystyrene	17.6 – 19.3	Poly(hexamethylene adipamide)	28.0
Poly(methyl methacrylate)	18.6 – 19.5		

Table 5.2. Solubility parameters for some common solvents

Solvent	Solubility parameter	Solvent	Solubility parameter
Perfluoroalkane	11.3 – 12.7	Acetone	20.3
n-Hexane	14.9	Cyclohexanone	20.3
Methyl cyclohexane	16.0	o-Dichlorobenzene	20.5
Cyclohexane	16.8	Nitrobenzene	20.5
Decalin	18.0	Dioxane	20.5
Toluene	18.2	m-Cresol	20.9
Ethyl acetate	18.6	Dimethylacetamide	22.1
Benzene	18.8	Cyclohexanol	23.3
Chloroform	19.0	Acetonitrile	24.3
Methyl ethyl ketone	19.0	Dimethylsulfoxide	24.6
Chlorobenzene	19.4	Dimethylformamide	24.8
Tetralin	19.4	Ethanol	26.0
Tetrahydrofuran	19.5	Methanol	29.7
Methylene chloride	19.8	Water	47.9
1,2-Dichloroethane	19.8		

solvents affects solute conformation and size. Some enzymes are unstable below pH 5 and DNA polymers are less soluble at low pH.

5.1.2 Sample Solutions

Sample concentration usually ranges from 0.1 to 0.5% depending on the MW of the sample. To dissolve the sample in a solvent may take between several hours to a day or so. Gentle shaking or warming at 50° to 60 °C is effective. Ultrasonic agitation should be avoided, because polymer samples, especially high-MW polymers, may degrade during ultrasonic treatment.

Sample solutions should be filtered manually through a 0.2 to 0.5 µm membrane filter in order to remove insoluble polymer gels and dust particles. THF should contain antioxidants, such as BHT (see Chap. 4), at concentrations between 0.025 and 0.05% to prevent the formation of peroxides. The antioxidant changes gradually and, therefore, it is wise to use fresh THF and to use sample solutions within 48 h.

5.2 Sample Concentration

5.2.1 Effects of Concentration

Sample concentration is one of the most important operating variables in SEC, because it is a known fact that the retention volumes of polymers increase with increased sample concentration. The concentration dependence of the retention volume is a well-known phenomenon and the magnitude of the peak shift to higher retention volume is more pronounced for polymers with higher MWs than those with lower MWs. It is observed even at a low concentration, such as 0.01%, although the peak shift is smaller than that at higher concentration. This effect is usually absent for polymers with a MW lower than 10^4.

In this sense, this concentration dependence of retention volume should be called the "concentration effect", not "overload effect" or "viscosity effect". If a large volume of a sample solution is injected, an appreciable shift in retention volume is observed even for low-MW polymers. This phenomenon is called the "overload effect". In the case of low-MW materials, if the sample concentration is too high, a peak shift to higher retention volume is sometimes observed. This phenomenon is also an "overload effect". An increase in the retention volume and significant band broadening is observed in the "overload effect", though band broadening may be insignificant in the case of the "concentration effect". The shift of retention volume caused by the flow nonuniformities in the system at higher concentrations (unusual use or preparative scale) may also be attributed to column overloading. If the solvent used as the mobile phase is viscous or the solution viscosity increases at high sample concentration, significant peak broadening due to viscous streaming, which gives rise to uneven flow in the columns, is observed on the rear side of the solute peak. This phenomenon is called "viscosity effect" or "viscous fingering".

An example for the concentration dependence of the retention volume is shown in Fig. 5.1. Retention volume increases linearly with increasing concentration of the sample solution and the magnitude of the increase is related to the increasing MW of the sample polymers. The extrapolation of the plots of retention volume vs. concentration to zero concentration gives the retention volume at infinite dilution (zero concentration).

The reason for the increase in retention volume with increasing polymer concentration is considered to be caused by the decrease in the hydrodynamic volume of the polymer molecule in the solution [1]. The hydrodynamic volume of a polymer is proportional to the MW of the polymer and, therefore, the

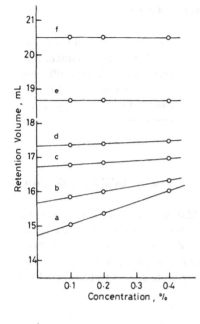

Fig. 5.1. Plots of retention volume vs. concentration for PS of different MW. Sample: PS of narrow MWD; mobile phase: THF; column: two Shodex KF 806L; injection volume: 0.1 ml; MW of sample: *a* 1.8 × 10⁶; *b* 6.7 × 10⁵; *c* 2.0 × 10⁵; *d* 97 200; *e* 20 400; *f* 2100

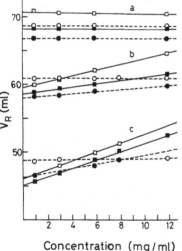

Fig. 5.2. Concentration dependence of retention volume for PS in good and theta solvents on silica gel columns. Mobile phase: benzene (□) and (■); benzene/methanol (77.8:22.2) (○) and (●); temperature: 25 °C (*open symbols*) and 60 °C (*filled symbols*); sample: *a* PS MW 51 000; *b* PS MW 1.6 × 10⁵; *c* PS MW 4.98 × 10⁵. Reprinted from Ref. [2] (© 1977) with kind permission from Elsevier Science Ltd, The Boulevard, Langform Lane, Kidlington 0X5 1GB, UK

retention volume is also a function of the hydrodynamic volume of the polymer. The hydrodynamic volume V_h is related to the polymer concentration c as

$$V_h = \frac{4\pi\,[\eta]\,M}{9.3 \times 10^{24} + 4\pi \times 6.022 \times 10^{23} \times c\,([\eta] - [\eta]_\theta)} \tag{5.2}$$

where $[\eta]$ and $[\eta]_\theta$ are the intrinsic viscosity of the polymer molecule in the solvent and in the θ-solvent, respectively. Equation (5.2) means that the concentration dependence of retention volume is negligible if a θ-solvent is used as the mobile phase.

An example is shown in Fig. 5.2 [2]. Benzene is a good solvent for PS and the retention volume increases linearly with an increase in sample concentration for PS having a MW higher than 5.1×10^4. A mixture of benzene/methanol (77.8 : 22.2, vol%) is a theta solvent for PS at 25°C and a dependence of retention volume on concentration is not observed. However, the thermodynamic quality of the solvent mixture is substantially improved at 60°C and the concentration dependence becomes distinct.

5.2.2 Influence of Concentration Dependence on MW Calculation

Retention volumes of PS standards of narrow-MW distributions (MWDs) at 0, 0.1, and 0.2% concentrations are plotted against the MW of the PS standards as shown in Fig. 5.3. This plot tells us that PS with a MW of 7×10^5 elutes at 16.0 ml when the sample concentration is 0.2%, whereas PS with a MW of 5.5×10^5 elutes at the same retention volume when the sample concentration is 0.1%.

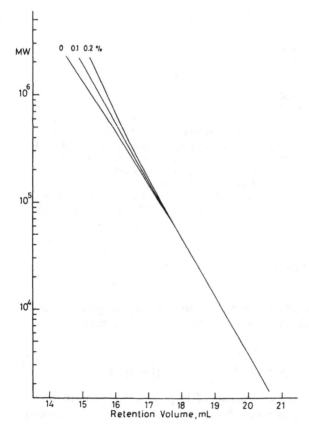

Fig. 5.3. Plots of MW vs. retention volume at several concentrations. Conditions are the same as those in Fig. 5.1. Numbers on the plots refer to concentration in %

Therefore, MW averages calculated with calibration curves of different concentrations may differ in value. Because the influence of sample concentration on the retention volume is based on the hydrodynamic volume of the polymer in solution, it is necessary to select experimental conditions that will reduce the errors produced by the concentration effect. The concentration of the injected sample solution starts to change immediately when it enters the column because of the MWD of the polymer sample. Therefore, even though the concentration of the polymer sample is the same as that of the polymer standards used for calibration, they elute from the column as if they were at different concentrations.

Table 5.3 shows the MW averages of PS NBS 706 (NIST, Washington, DC, USA) calculated with calibration curves of various concentrations [3]. Case 1 used calibration curves of the same concentrations as sample solutions, case 2 used a calibration curve obtained at 0.1% concentration, and case 3 used a calibration curve extrapolated to infinite dilution (0% concentration). The data presented in Table 5.3 demonstrate that calculated MW averages depend on the concentrations of the polymer standards for calibration, as well as sample concentrations. MW averages of PS NBS 706 calculated by using case 1 were higher than the vendor's values ($M_w = 257,800$, $M_n = 136,500$) and those using case 3 were lower, except for the 0.1% sample concentration. If the calibration curve at 0.1% concentration was used, reasonable values were obtained. Another procedure for obtaining reasonable values is to use calibration curves corresponding to concentrations of species at retention volumes, shown as case 4.

By rule of thumb, the preferred sample concentrations, if two SEC columns of 8 mm i.d. × 25 cm in length are used, are as follows. The sample concentra-

Table 5.3. MW averages of PS NBS 706 ($M_w = 257, 800$; $M_n = 136, 500$) measured at various concentrations

Case	Sample concentration (%)	Calibration concentration (%)	$M_w \times 10^{-5}$	$M_n \times 10^{-5}$
1	0.4	0.4	3.35	1.47
	0.2	0.2	3.01	1.43
	0.1	0.1	3.18	1.39
2	0.4	0.1	2.55	1.35
	0.2	0.1	2.71	1.38
3	0.4	0	2.30	1.23
	0.2	0	2.44	1.25
	0.1	0	2.84	1.26
4	0.4	Concentration of species at each retention volume	2.57	1.35
	0.2		2.55	1.31

Reprinted from Ref. [3] (© 1976) with kind permission from John Wiley & Sons, Inc., New York, NY, USA.

tion should be as low as possible and be 0.2% at most. For high-MW polymers, concentrations less than 0.1% are often required and for low-MW polymers, concentrations of more than 0.2% are possible. Samples with M_w over 10^6 must be diluted to a concentration of about 0.05%. The concentrations of PS standards for calibration should be one-half of the sample concentration. For PS standards with a MW over 10^6 it is preferable that they are one-eighth to one-tenth, and for those with a MW between 5×10^5 and 10^6, a quarter to one-fifth of the sample concentration. One possible guideline for the concentration of PS standards is as follows: PS MW over 10^6, less than 0.025 (w/v)%, those between 10^5 and 10^6, less than 0.05%, and others, less than 0.1%. When oligomers having an average MW below 10^4 are used for calculating MW, the problems concerning the concentration dependence are ignored.

Several calibration standards can be injected and analyzed at the same time, as long as all the peaks are baseline separated. However, it is preferabe that standards differ by a factor of about 10 in MW.

5.3 Calibration Curves

5.3.1 Calibration of Columns

In order to calculate the MW averages of a polymer from the SEC chromatogram, the relationship between the MW and the retention volume (the calibration curve) needs to be known, unless a light scattering detector is used. Details of approaches to calibration and MW-sensitive detectors are discussed in Chaps. 7 and 8, respectively. A calibration curve is a plot of log MW vs. retention volume of a column system and is used for the calibration of the MW of a solute eluted at a specified retention volume. The calibration curve is also used for the purpose of column selection (see Chap. 4).

In order to construct a calibration curve, several polymers and oligomers of known MW which have a narrow MWD should be used. For this purpose, polystyrene (PS), poly(methyl methacrylate), poly(ethylene oxide), pullulan, dextran, and sodium poly(styrene sulfonate) are now commercially available. The first three are used for non-aqueous SEC and the rest for aqueous SEC. Poly(ethylene oxide) is also used for aqueous SEC with columns packed with hydrophilic gels. The reason why polymer samples of narrow-MW distribution have to be used for calibration is that synthetic polymers have a MWD and the number-average MW (M_n) and the weight-average MW (M_w) of a polymer are not equal unless the polymer is truely monodisperse. Peak MW (M_p) of a polymer measured using an SEC system is not M_n or M_w but a value between them. If the polymer has a narrow-MW distribution, then M_p can be estimated as $(M_n + M_w)/2$ or $(M_n \times M_w)^{1/2}$, or sometimes M_w is assumed to be equal to M_n. The MWs of the calibration standards have to be determined by independent, absolute methods.

The MWDs of the calibration standards should be as narrow as possible and preferably the polydispersity M_w/M_n of the standards less than 1.05 for MW of standards from 2×10^3 to 1×10^6 and 1.20 for MW less than 2×10^3 and higher than 1×10^6. The peak asymmetry factor must lie in the range 1.00 ± 0.15. If the calibration standards give a shoulder on either side of the peak, the area represented by these anomalies should be less than 2.0% of the peak area. If not, the calibration standard should be rejected [4].

The calibration curve should cover the MW range of the polymer samples of interest. At least two calibration points must be measured per decade of MW and there must be at least five calibration points altogether. To obtain a graph of the calibration curve, the measured retention volumes (the ordinate, linear) are plotted against the corresponding values of M_p (the abscissa, logarithmic) on four- or five-cycle semilog graph paper. In the high-MW range, it is preferable that the peak of the first calibration standard eluted lies before the high-MW limit of the sample. If PS is used as the calibration standard, n-hexylbenzene ($M = 162$) can be used as the standard with the lowest MW on the calibration curve, because most PS standards are manufactured by anionically initiated polymerization and have a n-butyl group at one end.

5.3.2 Shape of the Calibration Curve

After plotting log MW vs. peak retention volume of polymer standards, a smooth curve is drawn through the calibration points to obtain a calibration curve. The calibration curve may be represented by a polynomial. The nth-order polynomial is given by

$$\log M = A + BV_R + CV_R^2 + DV_R^3 + \dots + EV_R^n \tag{5.3}$$

and

$$V = A' + B' (\log M) + C' (\log M)^2 + D' (\log M)^3 + \dots + E' (\log M)^n \tag{5.4}$$

where A, B, C, D, ..., E and A', B', C', D', ..., E' are the coefficients of the polynomial. These coefficients are determined by a least-squares fit of the polynomial to the experimental data points.

The third- to the seventh-order polynomials can be fitted, and currently the third-order polynomial is widely used. Although higher order equations improves the fit of the calibration curve, it can also lead to meaningless maxima and minima. Uneven orders have proved better, e.g. 3, 5, and 7 [4]. Sometimes the calibration curve is expressed as a linear function by

$$M = D_1 \exp(-D_2 V_R) \tag{5.5}$$

or

$$\log M = \log D_1' - D_2' V_R \tag{5.6}$$

where D_1, D_2 and D_1', D_2' are constants.

5.4 Column Selection

5.4.1 Non-aqueous SEC Columns

There are several columns commercially available. Among them, columns packed with PS gel are widely used for non-aqueous SEC. At the present time, SEC columns packed with PS gel of 5–10 µm particles are commonly offered by suppliers and are used rather than purchasing PS gels and packing them in the laboratory. The general aspects of columns and column packings are described in detail in Chap. 4. Column selection in SEC is based largely on the required MW range of separation and the choice of packing materials is limited.

The most important consideration in column selection is careful attention to the pore size of the packed gels, as this dictates the range of MW separation. However, the pore size of PS gels cannot be measured because of the semi-rigidity of the gel, and thus the MW at the exclusion limit is represented rather than the gel pore and is the important parameter for column selection. Improper selection of the pore size (MW at the exclusion limit) of a column leads to false peaks in SEC. For example, if a column has a MW at the exclusion limit of 10^5, then all molecules larger than MW 10^5 appear at V_0 and cannot be separated at all, even though the column has excellent resolution below MW 10^5 or a high number of theoretical plates. In this case, if all the molecules are larger than 10^5, then only one peak will be observed at the exclusion limit. When a portion of a polymer sample has a MW larger than 10^5, then this portion elutes at the exclusion limit and may show double peaks: one at the exclusion limit and the other after the exclusion limit. The presence of a peak at the exclusion limit is considered suspect in MW interpretation. A column with a larger exclusion limit must be used. Therefore, a column must be selected so that the highest MW of components in the sample is less than the MW at the exclusion limit. In other words, the resolution of a sample cannot be adjusted by changing the composition of the mobile phase, as in other LC modes, but is calculated from the calibration curve of the column.

For a better characterization of separation performance, the DIN Standard defines "separation performance" as [4]

$$\text{Separation performance} = \frac{(V_{R,M_x} - V_{R,(10M_x)})}{\text{Column cross-sectional area}} > 6.0 \qquad (5.7)$$

where V_{R,M_x} is the retention volume for PS of MW M_x in ml, $V_{R,(10M_x)}$ the retention volume for 10 times that MW in ml, and units of column cross-sectional area in cm^2. In general, to achieve an adequate separation, the calibration curve of log MW vs. retention volume for the column system used should not exceed a specified slope. This slope is measured using a pair of PS standards which elute in the area of the peak maximum for the polymer sample being analyzed or obtained from the calibration curve. The limit of 6.0 in Eq. (5.7) is chosen by experience. For a definition of resolution, see Chap. 2.

A series of Shodex KF columns is used by way of example to explain column selection (see Table 4.2 and Fig. 4.1). The MW at the exclusion limit for Shodex column KF 801 is 1600 as PS MW and this column can be used for the separation of oligomers and mixtures of small molecules with MW less than 1600. Shodex column KF 802 has an exclusion limit of MW 8000 and can be used for the separation of oligomers with MW less than 8000. Shodex columns KF 801, KF 802, and KF 802.5 are used for oligomer separation. The resolution for separating individual components in oligomers can be improved by coupling together two or more columns of the same pore size. While the range of separation remains the same, the volume over which separation is made is increased (see Chap. 2). For separating oligomers that extend over more than two columns, two or more columns of different pore sizes can be connected. Knowledge of the individual column calibration curves provides a useful guide for choosing which pair of pore sizes to use. However, excess connection of columns decreases the chromatogram response and causes considerable peak broadening.

Shodex columns KF 803 to KF 806 are used for polymer separation to calculate MW averages. Normally these four columns are connected in series, KF 806 first and KF 803 last. However, in order to keep high resolution (less peak broadening) and better chromatogram response, connection of two mixed gel columns of KF 806M is used in practice instead of using the four different columns. Shodex KF 806M is packed with PS gels of different pore sizes from KF 803 to KF 806. For the separation of polymers and calibration of MW averages, a broad range of separation and maximum linearity are required to yield accurate calculated values of MW averages. A broad range of separation is accomplished using KF 806M, for example. For both a broad range of separation and maximum linearity, the linear column KF 806L is preferable to KF 806M.

High-performance SEC (HPSEC) columns have a high number of theoretical plates such as 15000 per column (mostly 8 mm i.d. × 25 cm in length) and two columns are usually connected so that the total number of theoretical plates is 25000–30000. These values should be high enough for HPSEC. However, the DIN standard [4] requires a separation performance over 6.0 and most "linear" and "mixed" columns with an extended separation range from $M = 100$ to a few millions lie below 3.0 per column (25 cm in length). Therefore, connection of three columns may be needed to fulfil this requirement.

It is practically impossible to use columns with the same quality of packing in different laboratories. To obtain results that agree, as well as to reduce interlaboratory variations among different SEC instruments, it is necessary to adhere to the following minimum requirements: the number of theoretical plates, the resolution or the separation performance, types of columns and detectors, and so on.

5.4.2 Aqueous SEC Columns

For aqueous SEC, three types of gels are used: semirigid hydrophilic polymer gels, hydrophilic-group-bonded silica gel, and soft hydrophilic polymer gels.

Soft gels are not used for HPSEC. For details of columns, see Tables 4.3 and 4.4. Selection rules for aqueous SEC columns are the same as those for non-aqueous SEC. Mobile phases used for aqueous SEC are water or aqueous solutions containing neutral salts or buffers.

Polymer gels can be used for the separation of synthetic water-soluble neutral polymers, synthetic ionic polymers and natural polymers. When analyzing ionic polymers, the stationary phase must be neutral polymer gels or those having the same ionic groups, e.g. for poly(styrene sulfonate), cation-exchange resin can be used, but the use of an anion-exchange resin would adsorb the polymer. Pullulan, poly(ethylene oxide), and poly(styrene sulfonate) standards can be used for calibration.

Hydrophilic-group-bonded silica gels are used for the separation of proteins and ionic and non-ionic natural polymers. Pullulan, dextran, and several other protein standards are used for calibration. Poly(ethylene oxide) may not be applicable for calibration because it may show tailing due to residual silanol groups on the stationary phase.

5.5 Mobile Phase Selection

5.5.1 Non-aqueous SEC

Solvents that can be used for the mobile phase are limited to those that can dissolve sample polymers. Besides this limitation, when a PS gel is the stationary phase, the mobile phase must, as a rule, be a solvent which can dissolve non-cross-linked PS. These solvents are compatible with the PS gel and act to suppress the adsorption of the polymer samples to the surface of the PS gel. Although PS gel is cross-linked, PS gel swells somewhat in these solvents. Water, alcohols and acetone are nonsolvents for non-cross-linked PS and, as a result, contract the PS gel. Therefore, these solvents cannot be used. Although newer SEC columns may be tolerant to these polar solvents, the separation may no longer be based on molecular size with these polar mobile phases.

THF, chlorinated hydrocarbons (chloroform, methylene chloride, dichloroethane, etc.), aromatic hydrocarbons (benzene, toluene, o-dichlorobenzene and trichlorobenzene), dioxane, methyl ethyl ketone are good solvents for non-cross-linked PS and are compatible with PS gel. The degree of swelling of PS gels in these solvents should almost be the same and interchange of the solvent in the column with these solvents is possible. These solvents can be used for storing columns, although toluene is a better solvent for storage.

m-Cresol, hexafluoroisopropanol (HFIP), and dimethylformamide (DMF) are also used as mobile phases and can be used to replace the storage solvent in the column. However, the reverse replacement of these solvents to other solvents, such as THF, should be avoided. For small-pore columns in particular, the storage solvent should not be replaced with HFIP or DMF, because these solvents can shrink the swollen PS gel in the column. Although HFIP is not a solvent for PS, it can be used as the mobile phase for polyamide and poly(ethylene terephthalate); in this case, poly(methyl methacrylate) is used as the calibration standard instead of a PS.

The mobile phase should not be viscous at the temperature of separation. DMF and *m*-cresol are viscous at room temperature and thus operation at 50°–60 °C is preferable. The compatibility of the mobile phase with the detector is also important. If a differential refractometer is used as the detector, the refractive index (RI) of the mobile phase should be as different as possible from that of the sample. For instance, chloroform has a RI of 1.4457 at 20 °C which is nearly equal to that of diethylene glycol. Therefore, if poly(ethylene glycol) is the sample, the peak from diethylene glycol sometimes disappears or appears on the reverse side. Besides this problem, the response of each component is not proportional to its concentration, which will affect the accuracy of the MW averages. Similarly, if a UV detector is used, the mobile phase must have a lower absorption at the wavelength of detection.

Some solvents cause substantial changes in the calibration curves for PS columns. Figure 5.4 shows the effect of DMF with PS standards on PS gel columns [5]. When PS calibration standards are measured with DMF as the mobile phase, significant retardation is observed. However, the calibration curves of poly(ethylene glycol) standards with both THF and DMF mobile phases are almost the same.

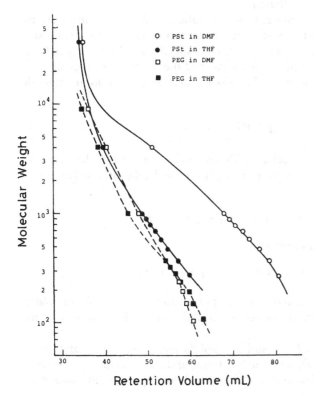

Fig. 5.4. Effect of mobile phase on calibration curves. Column (PS gel)/ mobile phase: eight Shodex GPC AD-802S (each 8 mm i.d. × 25 cm)/ DMF at 60 °C (*open symbols*); four Shodex GPC A-802 (each 8 mm i.d. × 50 cm)/THF at 23 °C (*filled symbols*). Reprinted from Ref. [5] (© 1980) with kind permission from Marcel Dekker Inc., New York, USA

Improper mobile phases collapse PS gel structures and destroy the column. Once swollen PS gel has shrunk in the column, restoration of the column performance is impossible. Drying the column by leaving it without a tight seal or injecting even a small amount of water or alcohols, which shrink the PS gel in the column, should be avoided.

THF containing a small amount (0.025–0.05%) of BHT as an antioxidant is recommended to prevent autoxidation and polymerization. Chloroform also contains about 1% ethanol as a stabilizer and should be used as such. THF is miscible with water and if it contains more than 1% water, then the elution of polymer samples tends to be retarded; the water content in THF should be kept at less than 0.05%.

5.5.2 Aqueous SEC

Water, aqueous solutions of neutral inorganic salts, such as sodium chloride and potassium chloride, and buffer solutions, such as phosphate buffer solutions, are used as mobile phases for most aqueous SEC. Aqueous solutions outside the pH range of about 2–8 degrade silica-based packings. Buffer solutions usually involve a mixture of a pair of weak acids and the salt of the weak acid, weak bases and the salt of the weak base, or a polybasic acid or base salt such as potassium acid phthalate.

Salt concentrations, pH values, types of buffers and their concentrations influence the retention volumes of ionic polymer samples. Neutral, acidic and basic inorganic salts are dissociated into their respective ions according to their dissociation constants. Therefore, the measure of the electrolyte concentration is not the molar concentration of the salts themselves, but the ionic strength μ, as follows:

$$\mu = 0.5 \sum_i c_i z_i^2 \tag{5.8}$$

where c_i is the molar concentration of ith ion and z_i is the unit of charge of the ion. The units of the ionic strength is mol/l. For example, the ionic strength of a 0.1 M Na_2SO_4 solution is

$$\mu = 0.5 \,(0.2 \times 1^2 + 0.1 \times 2^2) = 0.3 \text{ M}$$

and that of a 0.1 M NaCl solution is

$$\mu = 0.5 \,(0.2 \times 1^2 + 0.1 \times 1^2) = 0.1 \text{ M} \,.$$

With samples of non-ionic water-soluble polymers, SEC is not difficult. Water itself can be used as the mobile phase. Hydrophilic and ionic polymer gels are mostly used as packings.

Synthetic polymers, such as sodium poly(styrene sulfonate) and polyacrylamide, and natural polymers, such as fumic and alginic acids, have anionic groups in their chains. Stationary phases used for aqueous SEC also have residual anionic groups on the surface. Therefore, elution of these anionic poly-mers is affected by ion-exclusion effects. In order to suppress ion-exclusion, the addition of neutral salts to a water mobile phase or the use of buffer solutions as the mobile phase to control pH is required. However, an increase in the ionic strength may also cause an increase in adsorption. The addition of a small amount of organic solvent such, as ethanol or acetonitrile, may be effective to suppress adsorption.

The pH values of the mobile phase are very important for the separation of proteins. The ionic character of proteins changes from acidic to basic through neutral by decreasing pH values. For example, the isoelectric point of myo-globin is 7.1 and myoglobin acts as a neutral polymer at pH 7.1. Similarly, myoglobin acts as a cationic (basic) polymer at a pH value lower than 7.1, and as an anionic (acidic) polymer at a pH value higher than 7.1.

5.5.3 Preparation of the Mobile Phase

Reagent or extra-pure grade solvents can be used as mobile phases without further purification such as distillation. Recycling of solvents is also possible after puri-fication by simple distillation if the samples used are polymers. The mobile phase must be degassed and filtered before use, even if it has been distilled. Because solvents contain small particles of dust, they should be filtrated using a membrane filter of about 0.5 μm to protect the high-performance column from plugging and to prevent damage to the check valves of the solvent-delivery pump.

Dissolved air in the mobile phase may cause malfunctioning of the pump, flow rate variations, gas bubbles in the detector cell, and baseline fluctuations. The mobile phase should be degassed before it is introduced into the reservoir. Degassing can be carried out using a vacuum desiccator and an aspirator. The methods of degassing used are online purging with helium or vacuum degas-sing. The use of a vacuum degasser fitted between the reservoir and the pump is very effective and the device is now commercially available.

When columns used for aqueous SEC are to be stored or buffer solutions are to be used long term, the addition of a small amount of sodium azide (NaN_3) to the mobile phase is recommended as an antimicrobial agent to suppress micro-bial growth. An amount of 0.01–0.02% is usually sufficient.

5.6 Operating Conditions

5.6.1 Injection Volume

The retention volume of a polymer sample increases as the injection volume increases as shown in Fig. 5.5 [6]. At the same time, the number of theoretical plates N decreases as the injection volume increases due to band broadening. The increase in retention volume from an injection volume 0.1 to 0.25 ml was 0.65 ml, while that from 0.25 to 0.5 ml was 0.05 ml, suggesting that a precise or

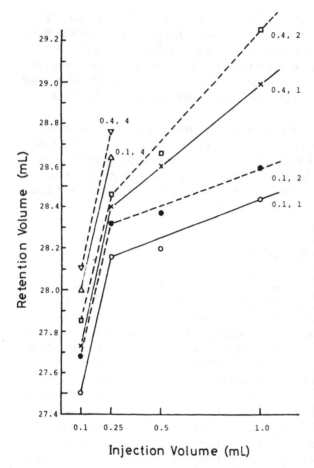

Fig. 5.5. Effet of injection volume on retention volume at different concentrations and flow rates. Numbers on curves refer to sample concentration in % and flow rate in ml/min, respectively; sample: PS MW 2×10^5; columns: four Waters µ-Styragel 10^6, 10^5, 10^4, and 10^3 Å. Reprinted from Ref. [6] (© 1977) with kind permission from John Wiley & Sons, Inc., New York, USA

constant injection volume is required if the volume is as small as 0.1 or 0.05 ml. In view of the significant effect of the injection volume on retention volume, it is important to use the same injection volume for the sample as that used when constructing the calibration curve. The use of a loop injector is essential and the same injection volume must be employed for all sample solutions including calibration standards regardless of MW values.

When selecting optimum experimental conditions, it must be considered whether it is more advantageous to inject a smaller volume of solution of higher concentration, or vice versa. From theoretical calculations, it appears that by increasing the concentration and injecting a lower volume to keep the total injected weight of the sample the same, the retention volume would increase. However, in most practical applications, this change is almost comparable with experimental errors. Injection conditions have to be chosen such that the total amount of the injected polymer is constant and the contribution to the total width of the chromatogram due to the injected volume is negligible within the limits of experimental error. The changes in the efficiency at equal sample load demonstrate that a lower concentration and a larger injected volume are more desirable than the other way around [6]. When two columns of 25 cm in length and 8 mm i.d. are used, a volume of 0.1 ml is adequate and a volume of 0.3 ml at most, considering detector stability.

The injection volume and the concentration of the sample solution should be matched to the total column volume. As a guide, the relationship between the injection volume and the sample concentration is as follows [7]:

Sample concentration (w/v%) =

0.05% × total column length (m)/the injection volume (ml)   (5.9)

For example, if the total column length is 1 m and the injection volume is 0.2 ml, then the sample concentration should be 0.125%. Taking this into account, the polymer concentration in the injected solution should not exceed 0.5 (w/v)% and the total injection volume 0.50 ml. For samples with M_w more than 10^6, the injection volume may be increased by two to three times and the concentration must be lowered (see Sect. 5.2.2). The injection volumes for the calibration standards should be coincident with those of the samples.

5.6.2 Flow Rate

Conflicting observations on flow rate dependence on retention volume have been reported, but recent data show that retention volume increases with incresing flow rate. These observations have been attributed to nonequilibrium effects, because polymer diffusion between the pores and interstices of gels is sufficiently slow that equilibrium cannot be attained at each point in the column. With decreasing flow rate, efficiency and resolution are increased. Bimodal distribution of a PS standard (NBS 705) with a narrow MWD distribution was clearly observed at a lower flow rate of 0.1 ml/min with a 8 mm i.d. column.

This dependence of retention volume on flow rate is not as important when selecting experimental conditions. Because most SEC experiments use a fixed flow rate suitable for the given column set, the flow rate dependence can be ignored. The most important contribution that flow rate makes to SEC is its influence on the efficiency and resolution of SEC columns. Figure 5.6 shows the effect of mobile phase flow rate on peak width for PS with MW of 2×10^5. The increase in peak width with increased flow rate is significant for polymers having a higher MW. Because PS samples have a MWD, even though the PS used in Fig. 5.6 had a narrow MWD, the observed peak width is the sum of the peak broadening due to the spreading of PS along the column and its MWD. The decreased peak width of PS with decreasing flow rate can, therefore, be regarded as caused by peak broadening within the column. A typical flow rate of 1 ml/min is used with 8 mm i.d. columns but a flow rate lower than 1 ml/min will give improved efficiency.

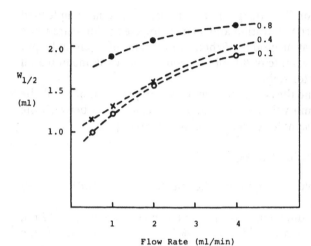

Fig. 5.6. Dependence of peak width at half-height on flow rate. Numbers on curves refer to sample concentration in %; injection volume: 0.1 ml; other conditions are the same as those in Fig. 5.5. Reprinted from Ref. [6] (© 1977) with kind permission from John Wiley & Sons, Inc., New York, USA

Very high-MW polymers are degraded at high flow rate, probably as a result of mechanical shear of large polymer chains. From this point of view, a lower flow rate is preferable. For samples with M_w over 10^6, a flow rate of 0.2 to 0.3 ml/min is preferable over 1 ml/min, or the particle sizes in the columns should be increased in the range of 20 to 30 µm.

However, a decrease in flow rate may cause larger errors in the calculated MW averages of sample polymers due to instrumental flow-rate fluctuations. This is because the separations are monitored according to retention time rather than retention volume. If the flow rate changes by 1 % between the time of sample injection and that of completion of the run (1 % flow rate drift), the errors of calculated MW averages amount to 10–15 %. Flow rate precision, defined as the standard deviation of the measured flow rate during operation and guaranteed by the manufacturer, is between 0.3 and 1.0 % (as relative standard deviation) at 1.0 ml/min and 0.002–0.003 ml at 0.1 ml/min. This means that if the flow rate is decreased to 0.1 ml/min, then errors in the calculated MW averages caused by flow fluctuation become serious. In particular, the flow-rate repeatability of the pump, defined as a standard deviation of the measured flow rate on a day-to-day basis under the same operating conditions, will be worse at lower flow rate. Therefore, a flow rate of 1 ml/min with a 8 mm i.d. column is often preferred as a compromise among resolution, speed, and accuracy of calculated MW averages.

According to DIN Standard [4], the reproducibility of retention volume measurement should be better than 0.3 %. The constancy and reproducibility of about 1 % that currently can be achieved over long operation times is inadequate for MW measurement. When the chromatograms are evaluated on the basis of retention time, it is necessary to check that the flow conditions during calibration and analysis are the same by using internal standards, a flow-rate meter, or manually by using a volumetric flask.

5.6.3 Column Temperature

For convenience of operation, SEC is normally carried out at room temperature, although column temperature can be increased to increase sample solubility or to decrease solvent viscosity. Polyolefins, such as polyethylene and polypropylene, require higher temperatures than 100 °C, because no solvent has been found that will dissolve these polymers at room temperature. SEC of proteins is preferrentially operated at room temperature to prevent denaturation of proteins at higher temperature.

Separation of molecules in SEC is governed mainly by the entropy change of the molecules between the mobile phase and the stationary phase and peak retention should be independent of temperature (see Chap. 2). However, an increase in retention volume with increasing column temperature is often observed. The relationship between the PS MW and the retention volume in the system PS gel/THF at column temperatures 15°, 25°, 35°, and 45 °C is shown in Fig. 5.7 [8]. A temperature difference of 10 °C results in a 10 to 15 % change in the MW. For example, the MW of PS eluted at 28 ml at 35° is 1.0×10^5, while the MW of PS eluted at the same retention volume at 25 °C is 1.15×10^5.

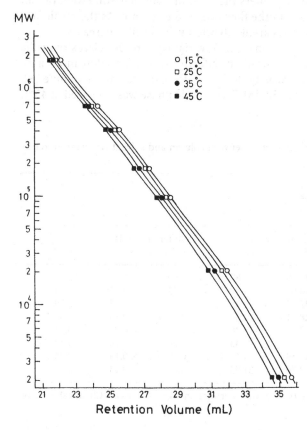

Fig. 5.7. Dependence of retention volume on column temperature for PS. Column: Shodex A80M PS gel; mobile phase: THF. Reprinted from Ref. [8] (© 1980) with kind permission from American Chemical Society, Washington, DC, USA

Two main factors that cause retention volume variations with column temperature are assumed: an expansion or a contraction of the mobile phase in the column and the secondary effects of the solute to the stationary phase. When the column temperature is 10 °C higher than room temperature, the mobile phase (temperature of the mobile phase is supposed to be the same as room temperature in this case) will expand about 1% from when it entered the columns, resulting in an increase in the real flow rate in the column due to the expansion of the mobile phase and a retention volume decrease. The magnitude of the retention volume dependence on solvent expansion is about one-half of the total change in the retention volume. The residual contribution to the change in retention volume is assumed to be that due to gel – solute interactions, such as adsorption. The magnitude of secondary effects increases with decreasing MW of solutes.

In order to obtain accurate and precise MW averages, the column temperature, as well as the difference of the temperatures in the solvent reservoir and the column oven, must be maintained. As can be seen in Table 5.4, if solvent (solvent reservoir) temperatures are different (25 °C (B) and 9 °C (C)), then retention volumes of the same solute differ from each other (B – C). When the column temperature (9 °C) is lower than the solvent (25 °C) before entering the pump, then the solvent, when it enters the column, contracts somewhat (about 1.5%), resulting in a decrease in the flow rate in the column. As the result, real retention time increases and apparent retention volume also increases.

Other factors affecting retention volume variations are the viscosity of the mobile phase, the size of gel pores, and the effective size of the solute molecules. Of these, the mobile phase viscosity does not affect the retention volume [8]. The results are shown in Table 5.5 [8]. The mobile phase was prepared at 45 °C

Table 5.4. Effect of temperature difference between column and solvent reservoir on retention volume

	Temperature (°C)					
Solvent	25	25	9			
Column	25	9	9			

	Retention volume (ml)			Difference (ml)		
	A	B	C	B – A	C – A	B – C
Benzene	40.30	41.20	40.87	0.90	0.57	0.33
PS 2100	35.30	36.02	35.77	0.72	0.47	0.25
PS 20400	31.60	32.27	31.98	0.67	0.38	0.29
PS 97200	28.30	28.81	28.50	0.51	0.20	0.31
PS 180000	27.05	27.63	27.30	0.58	0.25	0.33
PS 411000	25.25	25.75	25.43	0.50	0.18	0.32
PS 670000	23.92	24.39	24.11	0.47	0.19	0.28
PS 1800000	21.80	22.26	21.93	0.46	0.13	0.33

Reprinted from Ref. [8] (© 1980) with kind permission from American Chemical Society, Washington, DC, USA.

Table 5.5. Effect of solvent viscosity on retention volume of PS MW 670000

Mobile phase	THF	THF	5.2% PEG600/THF	5.8% PEG600/THF
Column temp (°C)	25	45	45	45
Mobile phase temp (°C)	25	25	25	25
		Retention volume (ml)		
	23.94	23.78	23.56	23.50
	23.98	23.50	23.56	23.52
	23.93	23.55	23.58	23.48
	23.94	23.49	23.53	23.42
	23.91	23.51	23.54	23.45
Average	23.94	23.51	23.55	23.47

Reprinted from Ref. [8] (© 1980) with kind permission from American Chemical Society, Washington, DC, USA.

by adding polyethylene glycol (PEG 600 (MW 600)) to THF so as to obtain the same viscosity as THF at 25 °C. A 5.2% PEG 600/THF solution at 45 °C had the same viscosity as THF at 25 °C. Similarly a 5.8% PEG 600/THF solution at 45 °C had the same viscosity as a 0.1% PS 670000 (MW 670000)/THF solution at 25 °C. PS having MW 670000 was used as a test sample. The retention volume of the test sample at 45 °C did not change in spite of the different viscosities of the

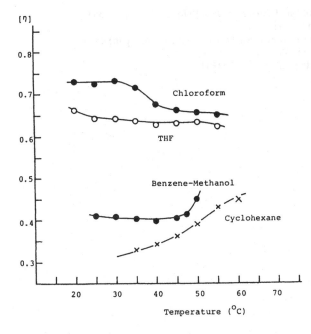

Fig. 5.8. Plots of intrinsic viscosity vs. concentration for unfractionated PS in different solvents. Reprinted from Ref. [9] (© 1984) with kind permission from Marcel Dekker Inc., New York, USA

mobile phase. These experimental results show the absence of a viscosity effect from factors affecting the retention volume.

A change in the size of gel pores with column temperature is small and improbable. The effective size of a solute molecule may change with changing column temperature. The dependence of intrinsic viscosity on column temperature for unfractionated PS in chloroform, THF, cyclohexane, and a mixture of benzene/methanol (77.8:22.2, vol%) is shown in Fig. 5.8 [9]. Temperature dependence of intrinsic viscosity of PS solutions was observed over a range of temperature. The intrinsic viscosity in THF was almost unchanged from 20° to 55 °C, while the intrinsic viscosity in chloroform decreased starting at 30° to 40 °C. Cyclohexane is a theta solvent for PS at around 35 °C and the intrinsic viscosity in cyclohexane increased with increasing column temperature. A mixture of benzene/methanol (77.8:22.2) is also a theta solvent at 25 °C and the intrinsic viscosity in the mixed solvent was unchanged up to 45 °C, but increased after 45 °C. Because the hydrodynamic volume is proportional to molecular size, the intrinsic viscosity can be used as a measure of the molecular size; column temperatures and solvents can be chosen where no changes in intrinsic viscosity are observed.

References

1. RUDIN A (1971) J Polym Sci Part A-1 9:2587
2. BLEHA T, BAKOS D, BEREK D (1977) Polymer 18:897
3. MORI S (1976) J Appl Polym Sci 20:2157
4. DIN 55672-1 Gelpermeationschromatographie (GPC) Teil 1: Tetrahydrofuran (THF) as Elutionsmittel (1995-02)
5. MATSUZAKI T, INOUE Y, OOKUBO T, MORI S (1980) J Liq Chromatogr 3:353
6. MORI S (1977) J Appl Polym Sci 21:1921
7. MORI S et al. (1996) BUNSEKI KAGAKU (Analytical Chemistry in Japan) 45:95
8. MORI S, SUZUKI T (1980) Anal Chem 52:1625
9. MORI S, SUZUKI M (1984) J Liq Chromatogr 7:1841

6 Molecular Weight Averages and Distribution

6.1 Calculation of MW Averages

6.1.1 MW Equations

Since a synthetic polymer is an assembly of molecules having different molecular weights (MW), the MW value has to be expressed as an average, and several different MW averages are defined (see Chap. 1). These MW averages are measured by different chemical and physical methods and are also calculated statistically in SEC by different equations. Consider a mixture of ten molecules with MW 100 and five molecules with MW 1000, then the average values can be calculated as follows. The arithmetic mean is the sum of the MW of all the molecules divided by the number of molecules as

$$M_n = \frac{100 + 100 + \dots + 100 + 1000 + \dots + 1000}{10 + 5} = \frac{10 \times 100 + 5 \times 1000}{15} = 400$$

The weighted mean is the sum of the weight times the MW of all the molecules divided by the total weight of the mixture as

$$M_w = \frac{100 \times 100 + 100 \times 100 + \dots + 100 \times 100 + 1000 \times 1000 + \dots + 1000 \times 1000}{100 + 100 + \dots + 100 + 1000 + \dots + 1000}$$

$$= \frac{10(100 \times 100) + 5(1000 \times 1000)}{10 \times 100 + 5 \times 1000} = 850$$

The former corresponds to the number-average MW, M_n, and the latter is defined as the weight-average MW, M_w.

There are five MW averages defined as

$$M_n = \frac{\sum w_i}{\sum N_i} = \frac{\sum (N_i M_i)}{\sum N_i} \tag{6.1}$$

$$M_w = \frac{\sum (w_i M_i)}{\sum w_i} = \frac{\sum (N_i M_i^2)}{\sum (N_i M_i)} \tag{6.2}$$

$$M_z = \frac{\sum (N_i M_i^3)}{\sum (N_i M_i^2)} \tag{6.3}$$

$$M_{z+1} = \frac{\sum (N_i M_i^4)}{\sum (N_i M_i^3)} \tag{6.4}$$

$$M_v = \left[\frac{\sum (N_i M_i^{a+1})}{\sum (N_i M_i)} \right]^{1/a} \tag{6.5}$$

where w_i is the weight of i molecules with MW M_i, N_i is the number of ith molecules with MW M_i, a is the exponent of the Mark–Houwink equation (see Chap. 7). M_z is defined as the z-average MW, M_{z+1} is the (z + 1)-average MW, and M_v is the viscosity average MW, respectively. M_n is related to the flexibility and tackiness of a polymer and is a function of the amount of low-MW material. M_w is related to the strength of the polymer as well as the content of high-MW chains in the polymer. M_z is related to of the brittleness and the amount of very high-MW material. M_v is related to the average solution viscosity.

By definition

$$M_n < M_v < M_w < M_z < M_z + 1 \tag{6.6}$$

If the polymer is a perfectly uniform (monodisperse) sample, then all the average MW values in Eq. (6.6) are equivalent.

The ratio of M_w and M_n

$$d = M_w/M_n \tag{6.7}$$

is the measure of the dispersity of the MWD in a polymer sample and is termed polydispersity. If d equals unity, then the polymer sample is a monodisperse sample. A number close to unity indicates that the polymer has a narrow-MW distribution.

General functions of MW averages from an SEC chromatogram are defined as

$$M_n = \frac{\int F(V) \, dV}{\int \{F(V)/M(V)\} \, dV} \tag{6.8}$$

$$M_w = \frac{\int F(V) M(V) \, dV}{\int F(V) \, dV} \tag{6.9}$$

where $F(V)$ is a function of an SEC elution curve and $M(V)$ is that of a calibration curve of the SEC column system. When the SEC chromatogram is normalized, then

$$\int F(V) \, dV = 1 \tag{6.10}$$

and Eqs. (6.8) and (6.9) can be simplified.

6.1.2 Calculation Procedure

A manual method to calculate MW averages is explained in this section using polystyrene (PS) as an example. PS SRM706 obtained from NIST (National

Institute of Standards and Technology, USA) is dissolved in tetrahydrofuran (THF) in a concentration of 0.2% and 0.25 ml of the solution is injected into an SEC system of two Shodex A-80M columns (50 cm × 8 mm i.d.). The mobile phase is THF and the flow rate is 1 ml/min. The detector is a refractometer and the attenuation is ×8. The SEC chromatogram of PS NBS 706 obtained under these conditions is shown in Fig. 6.1. A calibration curve is constructed with PS standards which are relatively monodisperse (see Chap. 7 for details of the calibration curve) and is shown in Fig. 6.1. The experimental conditions for the calibration curve should be the same as the measurement of PS NBS 706 except for the concentrations of PS samples, which are 0.1% for PS with MW less than 5×10^5 and 0.05% for those higher than 5×10^5.

The baseline of the SEC chromatogram must be drawn across the base of chromatogram from the starting point to the end point of the polymer chromatogram. Accurate baseline establishment is very important in order to obtain precise MW averages, especially the M_n value. A baseline can be manually drawn between the beginning of the polymer chromatogram and the end of the final solvent impurity peak after the polymer chromatogram. When the baseline is stable, the extrapolation of the baseline from the injection point of the sample solution to the end of the final solvent impurity peak agrees well with the base-

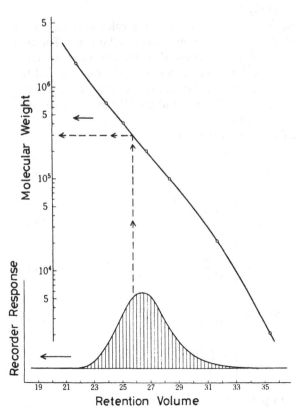

Fig. 6.1. SEC chromatogram of PS NBS 706 and a calibration curve. For experimental conditions, see text

line drawn between the beginning of the polymer chromatogram and the end of the final solvent impurity peak. These phenomena are often encountered when the experimental conditions are adequate and a stable detector is employed. If low-MW material is not included in the polymer sample, then the baseline is drawn to the end of the polymer distribution, as is the case in Fig. 6.1. If low-MW material is included in the polymer sample, the end of the polymer chromatogram is still above the baseline and is combined with solvent impurity peaks. The estimation of the end point for calculation is sometimes difficult in this case and an accurate calculation of M_n is not easy [1]. Similarly, if the baseline is not stable, precise and accurate MW averages cannot be obtained.

Manual digitization of the chromatogram is accomplished by dividing the baseline into equally spaced retention volumes as shown in Fig. 6.1. The chromatogram height at each spaced retention volume from the baseline to the chromatogram intersect is measured accurately (to about 0.1 mm scale). The number of heights determined at equally spaced retention volumes (the number of the divided points at the baseline) must be greater than 25 and that for PS NBS 706 in this case is 40–50, and ml is usually divided into quarters. For broader peaks, more divisions are required, while narrow peaks require less dividing points. The higher the number of incremental peak heights, the more accurate the MW averages are. If a computer is used for data acquisition, then the sampling frequency is normally 1 s.

Data for retention volume, height, MW and resulting calculated values are tabulated for the calculation of MW averages as shown in Table 6.1. The first column is the retention volume of the ith point V_i of the chromatogram, the second column is the height from the baseline to the chromatogram, and the third column is the MW obtained from the calibration curve at the ith point. For example, the height at $V_i = 25.75$ ml is 48.0 mm and the MW at the same point is read as 3.0×10^5 as shown in Fig. 6.1.

MW averages can be calculated according to the following equations:

$$M_n = \frac{\sum H_i}{\sum (H_i/M_i)} \tag{6.11}$$

$$M_w = \frac{\sum (H_i M_i)}{\sum H_i} \tag{6.12}$$

$$M_z = \frac{\sum (H_i M_i^2)}{\sum (H_i M_i)} \tag{6.13}$$

$$M_{z+1} = \frac{\sum (H_i M_i^3)}{\sum (H_i M_i^2)} \tag{6.14}$$

$$M_v = \left[\frac{\sum (H_i M_i^a)}{\sum (H_i)} \right]^{1/a} \tag{6.15}$$

where H_i is the height at the ith point of the retention volume.

Table 6.1. Data for calculation of MW averages of PS NBS 706

V_i (ml)	H_i (mm)	MW (M_i)	H_i/M_i	H_iM_i
21.5	0	2.06×10^6	0	0
.75	0.1	1.80×10^6	0.00000005	180000
22.0	0.3	1.60×10^6	0.00000019	480000
.25	0.7	1.42×10^6	0.00000049	994000
.5	1.7	1.27×10^6	0.00000134	2159000
.75	3.0	1.13×10^6	0.00000265	3390000
23.0	4.6	1.0×10^6	0.00000460	4600000
.25	7.2	9.0×10^5	0.00000800	6480000
.5	10.3	8.0×10^5	0.00001288	8240000
.75	14.1	7.0×10^5	0.00002014	9870000
24.0	18.1	6.3×10^5	0.00002873	11403000
.25	23.0	5.7×10^5	0.00004035	13110000
.5	26.7	5.1×10^5	0.00005235	13617000
.75	32.5	4.6×10^5	0.00007065	14950000
25.0	37.2	4.1×10^5	0.00009073	15252000
.25	41.5	3.7×10^5	0.00011216	15355000
.5	44.7	3.35×10^5	0.00013343	14974500
.75	48.0	3.0×10^5	0.00016000	14400000
26.0	50.0	2.7×10^5	0.00018519	13500000
.25	51.2	2.43×10^5	0.00021070	12441600
.5	51.4	2.18×10^5	0.00023578	11205200
.75	50.9	1.96×10^5	0.00025969	9976400
27.0	48.1	1.76×10^5	0.00027330	8465600
.25	45.0	1.58×10^5	0.00028481	7110000
.5	41.1	1.42×10^5	0.00028944	5836200
.75	36.8	1.28×10^5	0.00028750	4710400
28.0	32.0	1.14×10^5	0.00028070	3648000
.25	27.6	1.03×10^5	0.00026796	2842800
.5	23.7	9.3×10^4	0.00025484	2204100
.75	20.3	8.4×10^4	0.00024167	1705200
29.0	17.6	7.5×10^4	0.00023467	1320000
.25	15.0	6.7×10^4	0.00022381	1005000
.5	12.8	6.0×10^4	0.00021333	768000
.75	10.8	5.4×10^4	0.00020000	583200
30.0	8.8	4.8×10^4	0.00018333	422400
.25	7.2	4.3×10^4	0.00016744	309600
.5	6.0	3.8×10^4	0.00015789	228000
.75	5.0	3.35×10^4	0.00014925	167500
31.0	4.1	2.95×10^4	0.00013898	120950
.25	3.4	2.6×10^4	0.00013077	88400
.5	2.7	2.27×10^4	0.00011894	61290
.75	2.2	1.97×10^4	0.00011168	43340
32.0	2.0	1.70×10^4	0.00011764	34000
.25	1.4	1.48×10^4	0.00009459	20720
.5	1.0	1.27×10^4	0.00007874	12700
.75	0.9	1.08×10^4	0.00008333	9720
33.0	0.7	9.3×10^3	0.00007527	6510
.25	0.6	8.0×10^3	0.00007500	4800
.5	0.5	6.8×10^3	0.00007329	3400
.75	0.3	5.8×10^3	0.00005172	1740
34.0	0.1	4.8×10^3	0.00002083	480
.25	0	4.1×10^3	0	0
Sum	894.0		0.00651082	238311750

The height at the ith point H_i is divided by the MW at the ith point M_i and the values H_i/M_i are entered in the fourth column of Table 6.1. This value corresponds to the number of molecules at the ith point N_i. The product H_iM_i of the height and MW at the ith point is listed in the fifth column. The values from the second, the fourth and the fifth column are summed and MW averages are calculated using Eqs. (6.11) and (6.12) as

$$M_n = 894.9/0.00651082 = 1.37 \times 10^5$$

$$M_w = 238311750/894.9 = 2.66 \times 10^5$$

$$d = 2.66 \times 10^5/1.374 \times 10^5 = 1.94$$

Manufacturer's data for PS NBS 706 are $M_n = 1.37 \times 10^5$, $M_w = 2.58 \times 10^5$ (light scattering) or 2.80×10^5 (sedimentation). Other MW averages can be calculated similarly using Eqs. (6.13) to (6.15).

When the calibration curve is linear, then Eqs. (6.11) and (6.12) can be expressed as

$$M_n = \frac{D_1 \sum H_i}{\sum [H_i \exp(D_2 V_i)]} \tag{6.16}$$

$$M_w = \frac{D_1 \sum [H_i \exp(-D_2 V_i)]}{\sum H_i} \tag{6.17}$$

For definitions of D_1 and D_2, see Chap. 7. M_z, M_{z+1}, and M_v can also be derived from these equations. The calibration curve is usually slightly bent, though a linear column such as Shodex KF 806L (see Fig. 4.2) can be used, and polynomial representation is better in most cases. The calibration curve shown in Fig. 6.1 is presented as a 3rd-order polynomial as

$$\log M_i = 18.288 - 1.1154 V_i + 0.035936 V_i^2 - 0.00046319 V_i^3$$

The values for M_i obtained directly from the calibration curve as in Fig. 6.1 are rather inaccurate. Approximation of the calibration curve to a linear or a 3rd-order polynomial is more precise and accurate than by visual reading.

6.1.3 Reproducibility of Results

Careful measurement of SEC chromatograms gives precise MW averages within a relative error of 2–3%. However, large variations in results between laboratories, and even within a single laboratory, are observed routinely when columns or calibration standards have been changed. Round-robin experiments with the same sample polymers and calibration standards resulted in a relative standard deviation in M_w of about 5% and in M_n of 12% [2]. Several factors influencing the variations in calculated MW averages are discussed in more detail in

Chap. 5. In this section, the problems of the fluctuations of flow rate and on measuring the chromatogram height are explained.

(a) Influence of Flow-Rate Fluctuations. As explained above, the retention volume V_i corresponds to the molecular weight M_i, and MW averages are calculated using height H_i at retention volume V_i of the chromatogram and the corresponding MW at the same retention volume. Therefore, the strict correspondence of height-to-MW is essential to obtain accurate and precise MW averages and MWDs. Among several factors influencing the variations in the relationship between retention volume and MW, the fluctuations in flow rate by the pump is the greatest. When the fluctuation of the flow rate during the measurement is 1%, then the variation in MW is 15–20%. Therefore, it is important to use a high-performance pump to deliver the mobile phase as explained in Chap. 3.

Procedures to minimize the influence of flow-rate fluctuations are the measurement of flow rate and the use of an internal marker. The manual measurement of flow rate is simple, and an accurate and precise determination of MW averages can be expected. The effluent from the outlet of the detector is introduced into a 10-ml measuring flask through a PTFE tube and the time it takes to fill the flask with the mobile phase is measured. It is possible to measure the time to an accuracy of 0.1%. The retention volumes in Table 6.1 are apparent ones, because chromatograms are monitored according to retention time rather than retention volume. The abscissa of the chromatograms is not the retention volume scale, but the retention time scale and the value of the retention time multiplied by flow rate is used as the retention volume. The flow rate used in this multiplication is also an apparent one, because the value of the flow rate is not the measured one, but the setting value of the pump. By measuring the flow rate manually, true values of retention volumes on both the chromatogram and a calibration curve can be obtained. Several flow meters are now commercially available. Some are good and others not good enough for this purpose because of their low precision.

Internal markers are used for adjusting flow-rate variations and are usually low-MW compounds which should not interfere with the polymer peak. An internal marker is run with the polymer standards used for the construction of the calibration curve and the relative retention time of the standards to that of the internal marker is used instead of the retention volume (or retention time). When samples are run with the internal marker, the relative retention times of the samples to that of the internal marker are used to calculate MW averages.

One example of the effect of the use of an internal marker is that the within-day precision and the day-to-day precision of M_n and M_w were 2 and 0.8% as the relative standard deviation, and the long-term precision (10 months) were 2.7 and 1.1%, respectively, by using elemental sulfur as the internal marker [3]. Toluene, acetone, acetonitrile, sulfur, and other low-MW compounds can be used as internal markers. Some care is required when these low-MW compounds are used as internal markers, as these molecules shift much more than polymers as a consequence of non-size exclusion effects.

(b) Contribution of Measured Heights to Calculated MW Averages. M_n is calculated using the sum of H_i and the sum of H_i/M_i, and M_w using the sum of H_i and the sum of H_iM_i. The increase or decrease in these sums influences the values of calculated M_n and M_w. It is now worth investigating the extent of the contribution of H_i, H_i/M_i, and H_iM_i to their respective sums.

In Table 6.1, the height at $V_i = 32.0$ ml is 2 mm and the ratio of H_i/M_i to $\sum(H_i/M_i)$ at $V_i = 32.0$ ml is 1.5%. The other retention volume where the same ratio is obtained is 25.0 ml. The height at $V_i = 25.0$ ml is 37.2 mm. This means that when M_n is calculated, it becomes important to measure heights at the rear half of the chromatogram (the points of lower MWs) accurately, especially heights at the vicinity of the end of the chromatogram. However, the ratio of H_iM_i to $\sum(H_i/M_i)$ at 32.0 ml is 0.014% and that at 25.0 ml is 6.4%. Therefore, the extent of the contribution of the height at 32.0 ml is negligible, but that at 25.0 ml is large.

Similarly, the height at $V_i = 22.5$ ml is 1.7 mm and the ratio of H_iM_i to $\sum(H_iM_i)$ at $V_i = 22.5$ ml is 0.91%. The other retention volume where the ratio of H_iM_i to $\sum(H_iM_i)$ is the same is $V_i = 28.5$ ml. The height at $V_i = 28.5$ ml is 23.7 mm and the ratio of H_iM_i to $\sum(H_iM_i)$ is 0.92%, respectively. This observation indicates the necessity of the accurate measurement of heights at the first half of the chromatogram (the points of higher MWs) for the calculation of M_w, especially heights at the beginning of the chromatogram. The ratio of H_i/M_i to $\sum(H_i/M_i)$ at 22.5 ml is 0.02% and that at 28.5 ml is 4.1%. The extent of the contribution of the height at 22.5 ml is small, but that at 28.5 ml is large in the calculation of M_n.

These results suggest that the drawing of the baseline of the chromatogram correctly is more important than the digitization of the chromatogram. The heights at the starting point of the chromatogram should be measured precisely for M_w and those at the end point for M_n. When the baseline is smooth and without drift, that is, a flat, linear baseline prior to the emergence of the chromatogram is obtained and the chromatogram returns to that baseline after the peak has emerged, the heights at both ends of the chromatogram can be measured accurately. If the baseline drifts, it becomes impossible to accurately specify the starting and end points of the chromatogram.

The definition of the starting point of the digitization of the chromatogram is, in general, straightforward, since the baseline is usually flat and is not influenced by impurities or other sources at this point. Definition of the end point of the chromatogram is, however, normally difficult and inaccurate. When the polymer peak and peaks of low-MW materials such as additives, are separated adequately, it is easy to define the end point, but when the baseline resolution of these peaks is not attained, or when oligomer components appear following polymer components, it is not easy to establish the baseline. In these cases, the flat, linear baseline prior to the emergence of the chromatogram should be extended to the end of the impurity peaks or to the flat, linear baseline which appears after the impurity peaks have returned to baseline. After drawing the baseline, the end point has to be selected. One possible method of selecting the baseline, is to select the end point corresponding to MW 500 or 1000, depending on the content of additives and oligomers, and on the MW of the polymer [1]. One way to eliminate these problems is to connect a small column of narrow pore gel, such as Shodex KF-800D (100 × 8 mm i.d.), before the separation column(s) [4].

It is obvious that, when the baseline drifts, the calculated MW averages are not accurate, even though the flat baseline is defined smoothly. The value M_w can be calculated more accurately and precisely than M_n.

6.2 Calculation of MW Distribution

6.2.1 Equations

There are two types of molecular weight distribution (MWD): a differential MWD and a cumulative (or an integral) MWD. A differential MWD is illustrated by plotting the weight w_i or the weight fraction W_i of a polymer of a given MW M_i against M_i or $\log M_i$. When the ordinate is expressed as $d W/d (\log M)$ and the abscissa as $\log M$, then the total area of the differential MWD is normalized to unity as

$$\int W_{\log M} d (\log M) = 1 \qquad (6.18)$$

where $W_{\log M}$ is the weight fraction of molecules with molecular weight M. A cumulative MWD is illustrated by plotting the cumulative weight fraction or the % cumulative weight fraction on the ordinate and $\log M$ on the abscissa.

Theoretically derived MWD functions, which express the relationship between M and weight fraction $W(M)$, are: the logarithmic-normal distribution

$$W(M) = \frac{1}{\sigma \sqrt{2\pi}} \frac{1}{M} \exp \left[-\frac{1}{2\sigma^2} \left(\ln \frac{M}{M_o} \right)^2 \right] \qquad (6.19)$$

and the Schulz–Zimm distribution

$$W(M) = \frac{\lambda^{h+1}}{\Gamma(h+1)} M^h \exp(-\lambda M) \qquad (6.20)$$

In Eq. (6.19), σ is the width of the Gaussian distribution function and M_o is the MW at the peak apex of the distribution. The relations between these values and MW averages are as follows:

$$M_w = M_o \exp(\sigma^2/2) ; \qquad M_n = M_o \exp(-\sigma^2/2) ;$$

$$M_w/M_n = \exp \sigma^2 ; \qquad (M_w M_n)^{1/2} = M_o$$

In Eq. (6.20), Γ denotes the gamma function and h is the dispersion parameter, $h = [(M_w/M_n) - 1]^{-1}$, and $\lambda = h/M_n$. The relations between the parameters and MW averages are:

$$M_w = (h + 1)/\lambda \; ; \qquad M_n = h/\lambda \; ;$$

$$M_w/M_n = (h + 1)/h \; ; \qquad M_w = M_0$$

The true MWD calculated from an SEC chromatogram illustrates the shape of the MWD between these two functions and the MW at the peak apex also takes on the value between $(M_w M_n)^{1/2}$ and M_w. An SEC chromatogram exhibits a similar shape to the differential MWD, but both are different, because the former is expressed as the weight concentration of the sample or dW_i/dV_i against V_i, while the latter is presented as $dW_i/d(\log M_i)$ against $\log M_i$. Therefore, mathematical treatments must be made to represent the real differential MWD.

6.2.2 Calculation Procedure

(a) **Cumulative MWD.** A cumulative MWD is represented by plotting the % cumulative weight fraction (or cumulative weight fraction) CUM W_i vs. $\log M_i$. CUM W_i is defined as the weight fraction expressed as % up to the ith point from the end of the chromatogram and is different from the value of H_i/total height (%) (or simply $\sum W_i$). Table 6.2 illustrates the calculation procedure.

The first column is the retention volume of the ith point V_i as in Table 6.1. The second column is the sum of H_i from the lowest MW fraction $(i = n)$ up to the ith point and the third column is the percent of $\sum H_i$ divided by the total height. CUM W_i is calculated by adding one-half of H_i times 100 divided by the total height to the previously summed H_i up to the integer $i + 1$ from the lowest MW fraction $(i = n)$ times 100 divided by the total height as

$$\text{CUM } W_i(\%) = [(H_i/2)/\text{total height}]\,100 + \left(\sum_{i=n}^{i+1} H_i/\text{total height} \right) 100 \qquad (6.21)$$

and is listed in the fourth column. The integer i is counted from the largest retention volume (or the smallest MW) $(i = n)$ to $i = i + 1$ of the chromatogram. For example, CMU $W_i(\%)$ at $V_i = 27.0$ ml is

$$\text{CUM } W_i(\%) = [(48.1/2)/894.9]\,100 + (32.96/894.9)\,100 = 2.69 + 36.83 = 39.52\%$$

The total height equals the total area under the chromatogram and the total weight of the sample injected into the column system. CU $W_i(\%)$ divided by 100 is equal to the cumulative weight fraction and is equal to the weight fraction of the sample having a retention volume greater than V_i and MW less than M_i. It is desirable to represent the MWD on a semilogarithmic plot and CUM $W_i(\%)$ plotted against $\log M_i$ is the cumulative MWD, as shown in Fig. 6.2.

Table 6.2. Data for calculation of MWD of PS NBS 706

V_i (ml)	$\sum H_i$ (mm)	$\sum H_i$/total height (%)	CUM W_i (%)	dW_i/dV_i	$dV_i/$ $d(\log M_i)$	$dW_i/$ $d(\log M_i)$
21.5	894.9	100.00	100.00	0		0
.75	894.9	100.00	99.99	0.00045	4.667	0.0021
22.0	894.8	99.99	99.97	0.00134	4.832	0.0065
.25	894.5	99.96	99.92	0.00313	4.888	0.0153
.5	893.8	99.88	99.78	0.00760	4.958	0.0377
.75	892.1	99.69	99.52	0.01341	5.011	0.0672
23.0	889.1	99.35	99.09	0.02056	5.057	0.1042
.25	884.5	98.84	98.44	0.03218	5.118	0.1647
.5	877.3	98.03	97.46	0.04604	5.144	0.2376
.75	867.0	96.88	96.09	0.06302	5.205	0.3280
24.0	852.9	95.31	94.30	0.08090	5.229	0.4240
.25	834.8	93.28	92.00	0.10280	5.275	0.5423
.5	811.8	90.71	89.22	0.11934	5.288	0.6310
.75	785.1	87.73	85.91	0.14527	5.327	0.7739
25.0	752.6	84.10	82.02	0.16628	5.334	0.8891
.25	715.4	79.94	77.62	0.18550	5.362	0.9946
.5	673.9	75.30	72.81	0.19980	5.372	1.0733
.75	629.2	70.31	67.63	0.21455	5.376	1.1535
26.0	581.2	64.95	62.15	0.22349	5.376	1.2015
.25	531.2	59.36	56.50	0.22885	5.371	1.2291
.5	480.0	53.64	50.77	0.22975	5.361	1.2361
.75	428.6	47.89	45.05	0.22751	5.345	1.2161
27.0	377.7	42.21	39.52	0.21500	5.310	1.1449
.25	329.6	36.83	34.32	0.20114	5.301	1.0662
.5	284.6	31.80	29.51	0.18371	5.257	0.9658
.75	243.5	27.21	25.15	0.16449	5.237	0.8615
28.0	206.7	23.10	21.31	0.14303	5.186	0.7418
.25	174.7	19.52	17.98	0.12337	5.157	0.6362
.5	147.1	16.44	15.11	0.10593	5.111	0.5414
.75	123.4	13.79	12.66	0.09074	5.062	0.4593
29.0	103.1	11.52	10.54	0.07867	5.008	0.3940
.25	85.5	9.55	8.72	0.06705	4.952	0.3320
.5	70.5	7.88	7.16	0.05721	4.893	0.2799
.75	57.7	6.45	5.84	0.04827	4.831	0.2332
30.0	46.9	5.24	4.75	0.03933	4.767	0.1875
.25	38.1	4.26	3.86	0.03218	4.702	0.1513
.5	30.9	3.45	3.12	0.02682	4.631	0.1242
.75	24.9	2.78	2.50	0.02235	4.564	0.1020
31.0	19.9	2.22	1.99	0.01833	4.490	0.0823
.25	15.8	1.77	1.58	0.01520	4.421	0.0672
.5	12.4	1.39	1.23	0.01207	4.341	0.0524
.75	9.7	1.08	0.96	0.00983	4.273	0.0420
32.0	7.5	0.84	0.73	0.00894	4.195	0.0375
.25	5.5	0.61	0.54	0.00626	4.121	0.0258
.5	4.1	0.46	0.40	0.00447	4.049	0.0181
.75	3.1	0.35	0.30	0.00402	3.980	0.0160
33.0	2.2	0.25	0.21	0.00313	3.898	0.0122
.25	1.5	0.17	0.13	0.00268	3.806	0.0102
.5	0.9	0.10	0.07	0.00223	3.767	0.0084
.75	0.4	0.045	0.03	0.00134	3.657	0.0049
34.0	0.1	0.011	0.01	0.00045	3.556	0.0016
.25	0	0	0	0		0

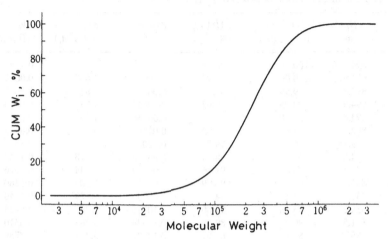

Fig. 6.2. Cumulative MWD from the data of Table 6.2 (PS NBS 706)

(b) Differential MWD. The height divided by the total height using a constant volume interval ΔV gives dW_i/dV_i in the fifth column in Table 6.2 as

$$dW_i/dV_i = H_i/\sum H_i/\Delta V \tag{6.22}$$

where dW_i is the weight fraction at the ith point. In Fig. 6.1 and Table 6.2, ΔV is 0.25 ml. The reason why the value H_i divided by the total height has to be divided by ΔV is that when normalization is performed by dividing H_i by the total height, the value becomes smaller with increasing number of dividing points. As a result, this decrease must be corrected by dividing the value by ΔV.

The next step is to obtain the reciprocal of the slope of the calibration curve and this is listed in the sixth column. If the equation of the calibration curve is already established, it is easy to obtain the reciprocal of the slope just by the differentiation of the equation. As a matter of fact, if the calibration curve is linear, the values of the reciprocal of the slope are all equal. The weight fraction per unit $\log M_i$ is calculated by multiplying the value in the fifth column at the ith point by the value in the sixth column of the same ith point and is listed in the seventh column of Table 6.2 as

$$\frac{dW_i}{d(\log M_i)} = \frac{dW_i}{dV_i} \frac{dV_i}{d(\log M_i)} \tag{6.23}$$

The differential MWD is plotted $dW_i/d(\log M_i)$ against $\log M$ as shown in Fig. 6.3.

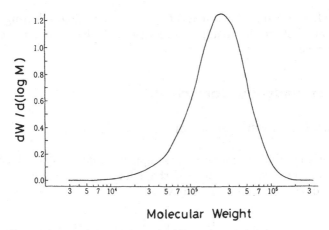

Fig. 6.3. Differential MWD from the data of Table 6.2 (PS NBS 706)

6.3 Automated Data Handling

Manual procedures for determining MW averages and MWD are time consuming. Even though a computer with in-house developed software can help the tedious calculations, there is also a high probability of operator errors.

Automated data handling systems save considerable time and lead to improvements in accuracy and precision by eliminating operator errors. Currently, many liquid chromatograph manufacturers sell data handling systems, some of which include software for SEC calculation. Most of them are real-time data acquisition and processing systems.

Data handling systems include an analog-to-digital (AD) converter coupled to an amplifier which converts analog signals to digital pulses. An SEC detector is connected to the data handling system and appropriate experimental conditions are put into the system. Input of data of peak positions of PS standards generates a linear or a polynomial function of a calibration curve. By injecting solutions of PS standards into the SEC system, calibration curve equations can be generated automatically.

After a sample solution has been injected into the SEC system, the operator chooses the beginning and end of data acquisition in order to eliminate taking extraneous baseline data and a baseline is generated automatically. Then data acquisition and processing starts and MW averages and MWD are calculated and printed out.

Some data handling systems have software that can determine the beginning and end of data acquisition automatically. As already mentioned in Sect. 6.1.4, determination of the beginning and end of the chromatogram is the most important factor for obtaining accurate and precise MW averages. For this reason, it is preferable that automated data handling systems have the following functions: after real-time data acquisition and processing, selecting the beginning and end of the chromatogram, redrawing the baseline, and recalculating MW averages.

6.4 Band Broadening Corrections

6.4.1 Band Broadening Equations

A concentration profile of a monomeric sample in a column is broadened when it passes through the column. The peak shape for such a monomeric species is approximated by a Gaussian function as

$$G(V) = \frac{1}{\sigma \sqrt{2\pi}} \exp\left[-\frac{1}{2\sigma^2}(V - V_0)^2\right] \tag{6.24}$$

where σ is the standard deviation of the Gaussian function, V is the retention volume variable, and V_0 is the retention volume of the chromatographic peak. Peak width at 0.607 height from the baseline is 2σ and that at half height is $2.354\,\sigma$. If band broadening is not observed, the peak shape becomes a rectangle. Band broadening occurs not only in the column but also in the connecting tubing, as well as the detector cell.

A synthetic polymer is, in general, a mixture of molecules with different MWs and has a MWD that has a similar shape to a Gaussian function. Thus, in an SEC chromatogram, band broadening due to the chromatographic processes superimposed on the peak which reflects the real MWD. The experimental SEC chromatogram $F(V)$ is expressed as a convolution between the true chromatogram $W(y)$ which is related to MWD and the Gaussian band broadening function $G(V)$. This convolution is described by

$$F(V) = \int W(y)\, G(V - y)\, dy \tag{6.25}$$

This equation is called Tung's equation and $G(V)$ is also termed the instrumental spreading function and is expressed as

$$G(V - y) = \frac{1}{\sigma \sqrt{2\pi}} \exp\left[-\frac{1}{2\sigma^2}(V - y)^2\right] \tag{6.26}$$

where V and y both represent retention volume and y is the retention volume of the chromatographic peak. The experimental SEC chromatogram is broader than the true chromatogram, owing to the instrumental band spreading and M_w calculated from $F(V)$ is larger and M_n is smaller than those from $W(y)$. The deconvolution problem is to solve $W(y)$ from Eq. (6.25).

6.4.2 Deconvolution Approaches

(a) **Solution by Polynomial.** One of the approaches for solving Eq. (6.25) is by using the polynomial representation of the chromatogram developed by Tung [5, 6]. The experimental and true chromatograms are described by

$$F(V) = \exp\left[-q^2 (V - y_o)^2\right] \sum_{i=0}^{n} S_i (V - y_o)^i \qquad (6.27)$$

and

$$W(y) = \exp\left[-p^2 (y - y_o)^2\right] \sum_{i=0}^{n} R_i (y - y_o)^i \qquad (6.28)$$

where q, y_o, and S_i are adjustable parameters and coefficients and p and R_i are an unknown parameter and coefficient to be determined, respectively. Values of p and R_i are related to q and S_i. First, values of q, y_o, and S_i are determined to fit $F(V)$ from Eq. (6.27) to the experimental chromatogram. Second, values of p and R_i are calculated from the values of q, y_o, and S_i and then calculate $W(y)$ from Eq. (6.28).

When n in Eq. (6.27) is 16, the chromatogram is indistinguishable from the curve representing the polynomial. When σ is not a constant with respect to the retention volume, $n = 4$ is used. To solve these equations, the value of σ is required. The instrumental spreading factor is defined as $h = 1/(2\sigma^2)$. When band broadening is approximated by the Gaussian function, the values of h can be determined by the reverse-flow technique [7]. A solution of PS with a narrow MWD is injected into the column and when the sample peak is approximately at the middle of the column, the direction of the flow is reversed and the concentration of the sample is monitored by a detector. The resulting chromatogram reflects only the peak formed by the band broadening processes. The broadening caused by the polydispersity of the sample has been completely cancelled by the flow reversal. This procedure is repeated by injecting a sample solution from the outlet of the column to measure the parameter h of the back half of the column. The overall h is then

$$h = 2/[(1/h_{\text{front}}) + (1/h_{\text{back}})] \qquad (6.29)$$

The parameter h varies with retention volume.

(b) **Solution by Iterative Methods [8].** Equation (6.25) can be transformed into Eq. (6.30) using the integral operator $G\{\ \}$ as

$$F(V) = G\{W(y)\} \quad \text{or} \quad F = G\{W\} \qquad (6.30)$$

$F(V)$ and other similar functions can be expressed simply as F and so on. The observed experimental chromatogram $F(V)$ is assumed to be the first approximation solution W_1. The corresponding uncorrected chromatogram $F_1(V)$ is calculated using Eq. (6.28) and the difference between $F(V)$ and $F_1(V)$ is obtained.

$$\Delta F_1 = F - G\{F\} \qquad (6.31)$$

The second approximation solution is assumed to be ΔF_1 and the calculation is repeated for ΔF_i

$$\Delta F_2 = \Delta F_1 - G\{\Delta F_1\}$$

$$\Delta F_i = \Delta F_{i-1} - G\{\Delta F_{i-1}\}$$

$$\Delta F_n = \Delta F_{n-1} - G\{\Delta F_{n-1}\}$$

By summing up these equations from $i = 1$ to $i = n$

$$F = \sum_{i=1}^{n} G\{\Delta F_{i-1}\} + \Delta F_n \tag{6.32}$$

where ΔF_0 equals F.

If $h_{\text{front}} = h_{\text{back}}$, then

$$\sum_{i=0}^{n} G\{\Delta F_i\} = G\left\{\sum_{i=0}^{n} \Delta F_i\right\}$$

and

$$F = G\left\{\sum_{i=0}^{n-1} \Delta F_i\right\} + \Delta F_n \tag{6.33}$$

By defining

$$W_i = \sum_{i=0}^{i} \Delta F_i$$

then

$$F = G\{W_{n-1}\} + \Delta F_n \tag{6.34}$$

When $n \to \infty$, ΔF_n approaches 0 and W is the solution of Eq. (6.25).

The second approach is to use the ratio of F and F_i instead of F_n. The first approximation is the same as the first approach. The second approximation solution W_2 is then given by

$$W_2' = \frac{F}{F_1} W_1$$

and

$$W_2 = \frac{W_2'}{\int W_2' dy} \tag{6.35}$$

The latter equation is simply a normalization step and a new F_2 is computed. The $(i + 1)$th approximation, W_{i+1}', is given by

$$W_{i+1}' = \frac{F}{F_i} W_i \tag{6.36}$$

This is equivalent to making a correction ΔW_i on W_i such that

$$\Delta W_i = \left(\frac{F - F_i}{F_i}\right) W_i \tag{6.37}$$

$$W_{i+1}' = W_i + \Delta W_i \tag{6.38}$$

When ΔW_i or the difference $F - F_i$ becomes less than is experimentally significant, W_i becomes the desired solution W.

6.4.3 Correction of MW Averages

The correction of peak broadening of the SEC chromatogram for obtaining accurate data of MW averages and MWD is a complex task and requires a computer. Instead of obtaining a real MWD, $W(V)$, before calculation of MW averages, correction of MW averages calculated from the experimental SEC chromatogram, $F(V)$, is simple and sufficient for the purpose of obtaining MW averages.

When SEC chromatograms are not skewed, that is, they show Gaussian distributions, the correction of peak broadening is simple [9]. Several PS standards of known MW are chromatographed and MW averages (M_n and M_w) (uncorrected for peak broadening) are calculated. These uncorrected average MWs are given the symbols $(M_n)_c$ and $(M_w)_c$. Assume that the reported values for the MW of these standards are the true values and assign them the symbols $(M_n)_t$ and $(M_w)_t$. Then calculate the quantity R as follows:

$$R = \left\{ \frac{(M_w)_t/(M_n)_t}{(M_w)_c/(M_n)_c} \right\}^{1/2} \tag{6.39}$$

The MW averages, $(M_n)_c$ and $(M_w)_c$, of sample polymers calculated from SEC chromatograms are corrected by

$$M_w = R(M_w)_c \quad \text{and} \quad M_n = (M_n)_c/R \tag{6.40}$$

If the SEC chromatograms are skewed, or could be skewed, then two correction factors are introduced [10]: Λ for symmetrical band-broadening peaks and sk for asymmetrical band-broadening peaks. A series of PS standards are chromatographed and MW averages are calculated. The quantities of Λ and sk are calculated by

$$\Lambda = \frac{1}{2} \left[\frac{(M_n)_t}{(M_n)_c} + \frac{(M_w)_c}{(M_w)_t} \right] \tag{6.41}$$

and

$$sk = (\phi - 1)/(\phi + 1) \tag{6.42}$$

where

$$\phi = \frac{(M_n)_t(M_w)_t}{(M_n)_c(M_w)_c} \tag{6.43}$$

The value Λ is always greater than unity and the sk value is usually greater than 0. These values are retention volume dependent and the values for correcting the sample MW averages are those which correspond to the retention volume of the sample peak. The sample MW averages are corrected by

$$M_n = (M_n)_c(1 + sk)\,\Lambda \tag{6.44}$$

and

$$M_w = (M_w)_c/(1 - sk)\,\Lambda \tag{6.45}$$

When a calibration curve is approximated by a linear function, then MW averages can be calculated with of band broadening correction by [11]

$$M_w = \exp\left[\frac{-(D_2\sigma)^2}{2}\right] \sum [F(V)\,D_1\exp(-D_2V)] \qquad (6.46)$$

and

$$M_n = \frac{\exp[(D_2\sigma)^2/2]}{[F(V)/D_1\exp(-D_2V)]} \qquad (6.47)$$

For D_1 and D_2, see Chap. 7. The term Λ is related to

$$\Lambda \simeq \exp(D_2\sigma)^2/2 \qquad (6.48)$$

References

1. Mori S, Kato H, Nishimura Y (1996) J Liq Chromatogr Rel Technol 19:2077
2. Mori S (1996) Bunseki Kagaku 45:95
3. Papazian LA, Murphy TD (1990) J Liq Chromatogr 13:25
4. Mori S, Marechal H, Suzuki H (1997) Int J Polym Anal Charact 4:87
5. Tung LH (1966) J Appl Polym Sci 10:375
6. Tung LH (1969) J Appl Polym Sci 13:775
7. Tung LH, Runyon JR (1969) J Appl Polym Sci 13:2397
8. Ishige T, Lee SI, Hamielec AE (1971) J Appl Polym Sci 15:1607
9. Hamielec AE, Ray WH (1969) J Appl Polym Sci 13:1319
10. Yau WW, Kirkland JJ, Bly DD (1979) Modern Size-Exclusion Liquid Chromatography, Wiley, New York, p 324
11. Yau WW, Stoklosa HJ, Bly DD (1977) J Appl Polym Sci 21:1911

7 Approaches to Molecular Weight Calibration

7.1 Calibration with Monodisperse Standards

A calibration curve is a plot of retention volume (abscissa) and log molecular weight (MW) (ordinate) of polymer solutes and is used to estimate the MW of a polymer eluted at a specified retention volume. One of the procedures to construct the calibration curve for a set of SEC columns is to use a series of narrow molecular weight distribution (MWD) standards of known MW and to relate the peak retention volume to the MW (peak position calibration curve). A sufficient number of narrow-MWD samples of the same polymer type, preferably 6 to 10 samples with different MW, are required as calibration standards.

Narrow-MW standards commercially available are polystyrene (PS) (Pressure Chemical Co., PA, USA; Toso Co., Tokyo, Japan; Polymer Laboratories Inc., UK; etc.) and poly(methyl methacrylate) (Polymer Laboratories) for non-aqueous SEC, and poly(ethylene oxide) (and glycol) (Toso), pullulan (linear polysaccharide) (Showa Denko Co., Tokyo, Japan), dextran (Pharmacia, Sweden), and sodium poly(styrene sulfonate) (Pressure Chemical) for aqueous SEC. A series of proteins with different MW, such as bovin serum albumin, ovalbumin, haemoglobin, lysozyme and cytochrome C, is also used for the peak position calibration.

The MW values of these standards are measured originally by membrane osmometry for M_n (number-average MW) and by light scattering for M_w (weight-average MW). Since these values are experimentally obtained by independent physical methods, the numerical values include errors and, therefore, calibration curves obtained from different combinations of the standards or from a series of standards from different sources (manufacturers) may not be in agreement, even though the same polymer type is used. For example, three different series of PS standards are available from three different manufacturers and thus three different calibration curves can be constructed for the same set of SEC columns. MW values obtained from these three calibration curves at the same retention volume are listed in Table 7.1. The difference between the maximum and the minimum MW at the same retention volume is 10 to 20% of the relative values. The use of a series of standards from the same source is therefore preferable.

The reason why narrow-MWD standards must be used to construct a calibration curve is as follows. As the narrow-MW standards available commercially are not true monodisperse polymers, the M_n and M_w of these polymers are not identical. This means that the peak retention volume cannot be related to M_n or M_w. Fortunately, the peak retention volume of the narrow-MWD polymer

Table 7.1. MW differences at the same retention volume due to different sources of PS standards

Retention volume (ml)	MW			Ratio of max MW/min MW
	A	B	C	
11	2450000	3000000	2500000	1.22
12	940000	1080000	900000	1.20
13	290000	340000	310000	1.17
14	86000	96000	100000	1.16
15	24500	26000	26000	1.06
16	6600	6100	6800	1.11

can be related to the mean of the two MWs, M_n and M_w, as $(M_n + M_w)/2$ or $(M_n M_w)^{1/2}$, or simply to M_w (not M_n). For a broad-MWD polymer, the peak retention volume cannot be related to these values.

A semilogarithmic chart is used to plot a set of data of peak retention volumes and MWs of a series of narrow-MW standards. The abscissa is a normal scale for retention volume and the ordinate is a logarithmic scale for MW. A smooth curve through the data points should be drawn. The third-order polynomial is usually approximated for the curve if needed as:

$$\log M = A + BV + CV^2 + DV^3 \tag{7.1}$$

To obtain a reciprocal of the slope of the calibration curve, Eq. (7.2) is convenient

$$V = A' + B' \log M + C' (\log M)^2 + D' (\log M)^3 \tag{7.2}$$

where A, B, C, D, and A', B', C', D' are constants, V is the retention volume, and M is the MW.

An example of a calibration curve is shown in Fig. 7.1. Experimental conditions are explained in Sect. 6.1.2. Values of MWs and retention volumes for a series of PS standards are: MW 2100 – V 35.3 ml; 20400 – 31.6; 97200 – 28.4; 200000 – 26.8; 411000 – 25.1; 670000 – 23.9; 1800000 – 21.8. The values of A, B, C, D are also shown in Sect. 6.1.2.

When a calibration curve can be approximated linearly, a linear function is generated as

$$M = D_1 \exp(-D_2 V) \tag{7.3}$$

or

$$\log M = C_1 - C_2 V \tag{7.4}$$

where D_1, D_2, C_1, and C_2 are constants and $-D_2$ corresponds to the slope of the calibration curve. C_1 and C_2 are the intercept and the slope of the calibration curve, respectively as they appear in the usual logarithmic calibration plot. The two sets of calibration constants are related as follows:

$$C_1 = \ln D_1 / \ln 10 ; \qquad C_2 = D_2 / \ln 10 \tag{7.5}$$

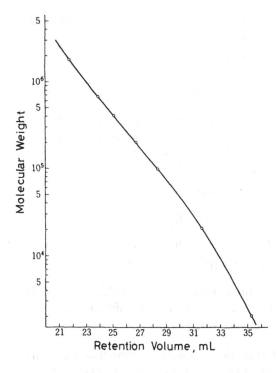

Fig. 7.1. Calibration curve obtained by PS standards. For experimental conditions, see Sect. 6.1.2

In Fig. 7.1, the range of retention volume 21.0 to 32.0 ml can be approximated to be linear, and a linear function of Eq. (7.4) is generated as $C_1 = 10.543$ and $C_2 = 0.1965$.

7.2 Calibration with Polydisperse Standards

7.2.1 Assumption of a Calibration Curve or MWD

Polymer types with narrow-MWD available for column calibration are limited and MW averages of polymers other than calibration standards are calculated merely as PS equivalent MW averages when a PS calibration curve is employed. If one or two polymer samples characterized by membrane osmometry and/ or light scattering are available, then the characterized polymers become the calibration standards and a calibration curve for the polymer concerned can be constructed.

(a) **Assumption of a Linear Calibration Curve.** When a calibration curve is assumed to be linear as expressed in Eq. (7.3), a linear calibration curve for the polymer concerned can be constructed using a polymer of known M_n an M_w, or two polymers with known M_n or M_w as follows. The goal is to find the values for D_1 and D_2 in Eq. (7.3), and then C_1 and C_2 in Eq. (7.4). First, the SEC chromatogram(s) of the polymer(s) concerned is measured and then MW averages of the polymer(s) are calculated using the following equations:

$$(M_w)_c = \sum F(V_i)\, D_1 \exp(-D_2 V) \tag{7.6}$$

and/or

$$(M_n)_c = 1/\sum (F(V_i)/D_1 \exp(-D_2 V) \tag{7.7}$$

where $F(V_i)$ is the SEC chromatogram and V_i is the retention volume at the ith point, respectively. The values of D_1 and D_2 are assumed and D_1 and D_2 are adjusted iteratively by a trial-and-error search until one of the values of F in the following equations is minimized [1]:

$$F_1 = \{(M_{w,1})_t - (M_{w,1})_c\}^2 + \{(M_{n,1})_t - (M_{n,1})_c\}^2 \tag{7.8}$$

$$F_2 = \{(M_{n,1})_t - (M_{n,1})_c\}^2 + \{(M_{n,2})_t - (M_{n,2})_c\}^2$$

$$F_3 = \{(M_{w,1})_t - (M_{w,1})_c\}^2 + \{(M_{w,2})_t - (M_{w,2})_c\}^2$$

$$F_4 = \{(M_{w,1})_t - (M_{w,1})_c\}^2 + \{(M_{n,2})_t - (M_{n,2})_c\}^2$$

The definitions of $(M)_t$ and $(M)_c$ are the same as described in Sect. 6.4.3. Suffix 1 and 2 means polymer 1 and polymer 2. The desired calibration curve is defined by the final values of the calibration constants.

(b) Assumption of a Nonlinear Calibration Curve. The calibration procedure using Eq. (7.8) can easily be modified to a nonlinear calibration curve such as a cubic equation. Four MW averages from two polymers with known M_n and M_w or four polymers with known M_n or M_w are required for the construction of the third-order nonlinear calibration curve.

When the SEC chromatogram of the polymer(s) of known MW averages is normalized, the chromatogram is expressed as Eq. (6.10) and the MW averages can be calculated from the SEC chromatogram and the calibration curve $M(V)$ of the polymer using Eqs. (6.8) and (6.9). If the calibration curve is expressed with the third-order polynomial as Eq. (7.1), the MW averages calculated are

$$(M_n)_c = U_n \exp A \tag{7.9}$$

and

$$U_n = 1/\int F(V)/\exp(BV + CV^2 + DV^3)\, dV$$

$$(M_w)_c = U_w \exp A \tag{7.10}$$

and

$$U_w = \int F(V) \exp(BV + CV^2 + DV^3)\, dV$$

The goal is to find the values of A, B, C, and D by a trial-and-error method. When two broad-MWD polymers of known M_n and M_w are available, then the value of F in the following equation should be minimized

$$F = \{(M_{w,1})_t - (M_{w,1})_c\}^2 + \{(M_{n,1})_t - (M_{n,1})_c\}^2$$

$$+ \{(M_{w,2})_t - (M_{w,2})_c\}^2 + \{(M_{n,2})_t - (M_{n,2})_c\}^2 \tag{7.11}$$

Trial values of B, C and D are assumed, the value of A is calculated from Eq. (7.12)

$$\exp A = \frac{U_n/(M_n)_t + U_w/(M_w)_t}{\{U_n/(M_n)_t\}^2 + \{U_w/(M_w)_t\}^2}$$ (7.12)

The calculations are repeated with other values of B, C and D until F is minimized [2].

(c) Assumption of MWD. The idealized MWD function for condensation polymers such as polyamides and polyesters can be given by the Flory "most probable distribution" function

$$W_r = (1 - P)\, r P^{r-1}$$ (7.13)

where W_r is the weight fraction of the polymer with a degree of polymerization r, P is the probability defined as $P = (M_n - M_0)/M_n$, and M_0 is the MW of the repeat unit. A plot of Eq. (7.13) is shown in Fig. 7.2a. The area under the curve is unity. The experimental SEC chromatogram for a condensation polymer is shown in Fig. 7.2b. The chromatogram has been normalized, so that the area under the curve is also unity [3].

The fractional area a_i (the shaded portion of the SEC chromatogram (Fig. 7.2b)) represents the weight fraction of the polymer that falls between the retention volume V_i and the end of the chromatogram. The same area a_i in Fig. 7.2a represents the weight fraction of the polymer that falls between the monomer and the degree of polymerization r_i. The MW at r_i is $r_i M_0$ which corresponds to the retention volume V_i. By making several such area comparisons over all the curves, a series of retention volume and MW pairs is generated and a calibration curve can be constructed for the polymer concerned.

If the polymer of interest is fractionated and the MW of each fraction is measured, then the cumulative (integral) MWD (MW vs. cumulative weight fraction) can be constructed. Similarly, the cumulative distribution of the retention volume vs. the cumulative weight fraction of the polymer of interest

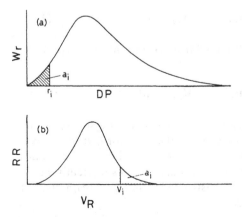

Fig. 7.2. a Most probable Flory distribution function and **b** SEC chromatogram of a condensation polymer. DP Degree of polymerization; RR recorder response; V_R retention volume; W_r is the polymer weight fraction. Reprinted from Ref. [3] (© 1972) with kind permission from John Wiley & Sons, Inc., New York, USA

can be constructed by measuring the SEC chromatogram as explained in Sect. 6.2.2. It is then possible to generate a series of retention volume – MW pairs by comparing both values of the cumulative weight fraction from these cumulative distributions. For example, polyethylene SRM 1475 supplied from NIST (USA) is a well-characterized linear polyethylene with a known cumulative MWD. Therefore, the calibration curve for linear polyethylene can be constructed using SRM 1475.

7.2.2 Use of Primary and Secondary Standards

The method decribed here is to use narrow-MWD standards as a primary standard and one or two polymers of known MW averages as the secondary standard [4]. Any assumption of a calibration function or MWD is not required.

Assume that the MW of polymer B, $(M_B)_i$, can be related to the MW of polymer A, $(M_A)_i$, eluting at the ith point of the retention volume by the expression

$$(M_B)_i = s\,(M_A)_i^t \tag{7.14}$$

where s and t represent constants to be calculated. Polymer A is the PS standard used to construct the primary calibration curve, if PS standards can be used in the SEC system, and polymer B is the secondary standard. Prepare one broad-MWD polymer (B) with known $(M_n)_t$ and $(M_w)_t$. Measure the SEC chromatogram of this secondary standard polymer B and calculate the MW averages using Eq. (7.14)

$$(M_n)_c = \sum H_i / \sum \{H_i / s\,(M_A)_i^t\} \tag{7.15}$$

and

$$(M_w)_c = \sum \{H_i\,s\,(M_A)_i^t\} / \sum H_i \tag{7.16}$$

The values of $(M_A)_i$ are given by the primary calibration curve.

Estimate the value of t first, then the MW averages of polymer B can be calculated as

$$(M_n)_c = k_1 s \quad \text{and} \quad (M_w)_c = k_2 s \tag{7.17}$$

where k_1 and k_2 are numerical values obtained from H_i and $(M_A)_i^t$ in Eqs. (7.15) and (7.16). Assume

$$(M_n)_t = (M_n)_c \quad \text{and} \quad (M_w)_t = (M_w)_c \tag{7.18}$$

then two values of s are obtained, which may not be equal in most cases at the first attempt. The goal of this calculation is to find the value for t where the two values of s from Eq. (7.17) are equal. The process is repeated with other values of t until the difference in the two values of s in Eq. (7.17) is sufficiently small. A calibration curve for polymer B can be constructed from Eq. (7.14), a final set of t and s values, and a PS calibration curve.

Besides one polymer and its M_n and M_w, two polymers with either two $(M_n)_t$, two $(M_w)_t$ or one $(M_n)_t$ and one $(M_w)_t$ can also be used as secondary standards. An example of a pair of s and t values for poly(vinyl chloride) (PVC) in tetrahydrofuran (THF) is 0.189 and 1.10, which was obtained using a PS calibration curve and two PVC standards. Because of peak broadening, preferable pairs are one polymer with known M_n and M_w or two polymers, given one M_n and one M_w. This procedure can be applied to systems where interactions between sample polymers and the stationary phase are observed, to which the universal calibration method cannot be applied.

The calibration curves constructed in Sect. 7.2 are effective for polymer samples having MW averages and a MWD similar to those used for calibration. The true calibration curves may be obtained by a peak position calibration approach. However, unlike those from narrow-MWD standards, although the calibration curves obtained by broad-MWD polymers are affected by band broadening, the calculation of MW averages using these latter calibration curves minimizes errors due to band broadening.

7.3 MW vs. Molecular Size

7.3.1 Terminology of Molecular Size

Although a calibration curve is a plot of log MW rather than log molecular size against retention volume, SEC separates polymers according to molecular size not MW. If the same type of polymer is used for calibration and samples then it is possible to explain SEC as separating polymers according to MW, because molecular size is proportional to MW.

There are several size parameters that can be used to describe the sizes of polymer molecules: e.g. radius, diameter, molar volume, and so on. The shape of a polymer in solution can be described as a spherical particle, a rigid cylindrical rod, or a flexible random coil; however, these size parameters are not always the best way to describe the sizes of polymers with different shapes. For example, even though the radii of a spherical particle and a cylindrical rod molecule are the same, these two molecules would not necessarily have the same molecular size. These size parameters can be applied only to describe polymers having the same molecular shape.

If polymer solutes elute exclusively by size exclusion without any interaction with the stationary phase, polymer solutes eluted at the same retention volume would have the same molecular size in solution, apart from the definition of molecular size. Therefore, a comparison of molecular size obtained by SEC and that measured by other methods is useful to estimate the real shape of polymer molecules in solution.

A synthetic polymer in solution usually exists as a random coil and the size of the molecular coil is influenced by the force of the polymer–solvent interaction. In a thermodynamically "good" solvent, where polymer–solvent contacts are highly favored, the coil is relatively extended. In a "poor" solvent, it is relatively contracted. This means that the molecular size varies with the solvent used to dissolve a polymer and the temperature of the solution. MW, however, is an invariable parameter to describe the dimensions of the polymer.

7.3.2 Parameters for Molecular Size

(a) **Hydrodynamic Volume.** Einstein's viscosity equation that expresses the relationship between the viscosity of a colloidal solution and the concentration is

$$\eta = \eta_0 \{1 + (bN\phi'/V')\} \tag{7.19}$$

where η is the viscosity of the solution, η_0 is the viscosity of the solvent, N is the number of dispersed particles (equal to the Avogadro number in the case of 1 mole of particles), ϕ' is the volume of each particle, V' is the volume of the solution, $N\phi'/V' (= \psi)$ is the volume fraction occupied by the dispersed particles in the solution, and b is a constant which varies with the shape of the particles.

There are four conditions needed to satisfy Eq. (7.19).

(1) The dispersed particle should be considerably large as compared to the molecular size of the solvent and the free diffusion distance.
(2) The dispersed particle should be a rigid body, its surface should be in contact with solvent molecules, and there should be no slipping between the particles.
(3) The movement of the dispersed particle is not impeded by other particles because of low concentration.
(4) Solvent molecules flow according to Hagen-Poiseuille's law.

Under these conditions, the viscosity equation for a dilute solution of rigid sphere particles is

$$\eta = \eta_0 (1 + 2.5\,\psi) \tag{7.20}$$

When the concentration is expressed as c (g/ml) and the specific gravity of the solute is d_s, then $\psi = c/d_s$ and $\eta_{sp} = 2.5\,c/d_s$ (η_{sp} is the specific viscosity), and Eq. (7.20) is expressed as

$$[\eta] = \lim_{c \to 0} \frac{\eta_{sp}}{c} = \frac{2.5}{d_s} = \frac{2.5\,N_A}{M/V_h} = K' \frac{V_h}{M} \tag{7.21}$$

where $[\eta]$ is the intrinsic viscosity (ml/g), V_h is the hydrodynamic volume, N_A is the Avogadro number, M is the MW and K' is a proportional constant.

In general, the hydrodynamic volume is expressed in terms of the product of the intrinsic viscosity and the molecular weight M of the polymer sample as $[\eta]M$, but in reality the value of the product $[\eta]M$ is directly proportional to the hydrodynamic volume as expressed in Eq. (7.21).

(b) Hydrodynamic Radius. The hydrodynamic radius R_h or the "equivalent hydrodynamic sphere" (see Sect. 7.3.3) of a random coil for a polymer chain can be expressed as

$$\eta_{sp}/c = 2.5\,N_A(4/3)\,\pi\,R_h^3/M \tag{7.22}$$

For a dilute solution

$$R_h = \{3M[\eta]/(10N_A\pi)\}^{1/3} \tag{7.23}$$

(c) Radius of Gyration. The mean radius of a macromolecule, which consists of many atoms or atomic groups regarded as mass points, is termed the radius of gyration R_g which is the root-mean-square distance of the elements in the chain from its center of gravity, as shown in Fig. 2. When the mass of a component unit i is m_i and the distance from the center of mass of the macromolecule to the ith component is S_i, then

$$R_g = (\textstyle\sum m_i S_i^2/\sum m_i)^{1/2} \tag{7.24}$$

For a homopolymer of n repeat units that consists of one type of monomer unit, all m_i elements are the same and Eq. (7.24) becomes

$$R_g = (\textstyle\sum S_i^2/n)^{1/2} \tag{7.25}$$

The value obtained by light scattering is the z-average mean-square radius of gyration $\langle S^2\rangle_z$ and for a true monodisperse polymer, $R_g = \langle S^2\rangle_z^{1/2}$. For a polydisperse polymer, the z-average radius of gyration is expressed as

$$\langle S^2\rangle_z = \frac{\sum(W_i M_i\langle S^2\rangle_i)}{\sum(W_i M_i)} \tag{7.26}$$

(d) End-to-end Distance. The mean distance between the ends of the chain for a random coil is termed the end-to-end distance (see Fig. 2.8). A model of a random coil for a polymer chain consists of a series of n links of length l joined in a linear sequence with no restrictions on the angles between successive bonds and the root-mean-square end-to-end distance is

$$\langle r^2\rangle^{1/2} = n^{1/2}l \tag{7.27}$$

where n is the number of segments in the polymer and l is the length of one segment.

(e) Effective Radius. The effective radius R_e of a linear polymer is defined as being equal to half of the mean maximum cross-section of the fluctuating random coil and expressed as

$$R_e = (\pi n l^2/6)^{1/2}/2 \tag{7.28}$$

(f) Stokes' Radius. When a spherical particle with a radius of R_s moves with a speed u through a stationary fluid having a viscosity, the force F produced by the fluid is

$$F = 6\pi R_s\eta u \tag{7.29}$$

This equation is known as Stokes' equation and R_s is defined as the Stokes' radius. The Stokes–Einstein equation is

$$D = kT/6\pi R_s \tag{7.30}$$

where D is the diffusion coefficient of the particle, k is the Boltzmann constant, and T is the absolute temperature. The relation between R_s and MW is

$$R_s = M^{1/3} V^{1/3} (f/f_o)^3/(4\pi N_A)^{1/3} \tag{7.31}$$

where V is the particle molar specific volume and f/f_o is the ratio of friction. When the molar volume of the solvated particle is V_s, that of nonsolvated particle is V_u, then $V_s/V_u = (f/f_o)^3$.

7.3.3 Comparison of Size Parameters

The relationship between the radius of gyration and the mean-square end-to-end distance is $R_g^2 = \langle r^2 \rangle / 6$. When the radius of a spherical particle is R_r, then $R_g^2 = (3/5) R_r^2$ and for a rigid cylindrical rod of length L, $R_g^2 = L^2/12$.

In a polymer solution, the solvent molecules in the random coil of the polymer move with the polymer molecules. This random coil attached to solvent molecules is called the equivalent hydrodynamic sphere. The relationship between the intrinsic viscosity $[\eta]$ and the MW for the equivalent hydrodynamic sphere is

$$[\eta] = \phi_o \langle r^2 \rangle^{3/2} M^{-1} \tag{7.32}$$

and

$$[\eta] = 6^{3/2} \phi_o R_g^3 M^{-1} \tag{7.33}$$

where ϕ_o is the Flory universal constant, equal to 2.1×10^{23} (theoretical) or 2.55×10^{23} (mol^{-1}) (experimental).

Equations (7.21), (7.23), (7.32) and (7.33) explain the relationship between the hydrodynamic volume and the molecular size, and the mutual relationship of each molecular size. The relationship between R_h and R_g is

$$R_h/R_g = 3.69 \times 10^{-7} \phi_o^{1/3} \tag{7.34}$$

The dependence of the size of polymer molecules on MW varies as a function of the conformation of the polymers. Table 7.2 lists several molecular sizes for polystyrene (PS) and proteins in relation to MW. The relationship between R_g and MW for PS is $R_g = 0.137 M^{0.589} (\text{Å})$ and between R_g and R_e is $R_e = 0.866 R_g$.

Table 7.2. Relationship between molecular weight and molecular size for polystyrene and proteins[a]

PS					Protein		
MW × 10⁻⁴	THF, 25°C R_g (Å)	R_e (Å)	TCB, 130°C R_h (Å)	THF, 23°C R_h (Å)	Name	MW × 10⁻⁴	R_S Å
0.125	9	9			Cytochrom C	1.34	16.5
0.225	13	12			Myoglobin	1.70	20.8
0.40	18	16			Chymotrypsino-	2.30	22.4
1.03	32	28			gen A		
1.98			34.4		Ovalbumin	4.35	27.3
2.04	47	42			Bovine serum		
5.10	81	72	57.5	61.0	albumin	6.70	35.5
9.72	119	105	83.5	88.3	Aldolase	14.80	46.0
16.0	159	141	111	118	Catalase	23.0	52.3
24.5			146		Ferritin	44.0	60.6
39.8			188	203	Thyroglobulin	67.0	86.0
41.1	277	241					
78			276	300			
86.0	428	379					
180.0	660	585					

[a] THF – tetrahydrofuran; TCB – 1,2,4 trichlorobenzene.

7.4 Extended Chain Length

7.4.1 Extended Chain Structure

The random coil of a synthetic polymer arises from the relative freedom of rotation associated with the chain bonds of most polymers and the large number of conformations accessible to the molecule. One of these conformations is the fully extended chain structure, and the contour length of the chain can be calculated in a straightforward way.

Figure 7.3 shows the fully extended chain structure for linear polyethylene. As the valence angle for a carbon–carbon chain is about 110°, the extended chain structure is not a straight line, but shows a planar zigzag carbon chain in the case of an all-trans form. The bond length between carbon atoms is 1.54 Å and the projected length in the direction of the chain axis becomes 1.26 Å, thus the projected length of a monomer unit in the chain direction is 2.52 Å. The total length is the extended chain length of the polymer.

Olefinic and vinyl polymers are composed of carbon–carbon chains as the polymer backbone and the extended chain structures for these polymers are similar to that shown in Fig. 7.3. Polypropylene, PS, and PVC, for example, replace a hydrogen atom (marked as H* in Fig. 7.3) with a methyl group, a phenyl group, and a chlorine atom, respectively. Poly(methyl methacrylate)

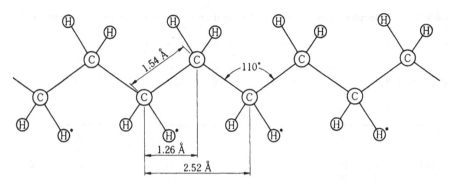

Fig. 7.3. Fully extended planar zigzag carbon – carbon chain for linear polyethylene

(PMMA) replaces two hydrogen atoms with a methyl group and a carboxy methyl ester.

A minimum repeating unit (a monomer unit) for these polymers is a two-carbon chain and polyethylene, for example, with a degree of polymerization n, is expressed as $-(CH_2-CH_2)_n-$. The extended chain length of the monomer unit is 2.52 Å and the extended chain length of the polymer with a degree of polymerization n is 2.52 n. For example, if $n = 1000$, the polyethylene has a MW 28000 and the extended chain length is 2520 Å. Similarly PS with $n = 1000$ has MW 41300 and a chain length 2520 Å. It should be stressed that the size of the random coil must be expressed in terms of the statistical parameters explained in Sect. 7.3, not the extended chain length.

7.4.2 Q-Factor

The Q-factor is obtained by dividing the MW of the repeating unit of the polymer (MW of the monomer unit) by the extended chain length of the repeating unit (the monomer unit) and is defined as the MW per 1 Å extended chain length. From this definition, the extended chain length of a polymer is obtained by dividing the MW of the polymer by the Q-factor of the polymer. The MW of the monomer unit of PS is 104 and thus the Q-factor of PS becomes 41.3 by dividing 104 by 2.52 Å. Q-factors of several other polymers are as follows: polyethylene – 11.1, polypropylene – 16.7, PVC – 24.8, PMMA – 39.7, poly(butyl methacrylate) – 56.3.

MW averages of a polymer other than PS calculated using a PS calibration curve are not the real MW averages of the polymer, but the PS equivalent MW averages which can be converted to real MW averages of the polymer concerned using the Q-factor as follows:

$$MW = PS \text{ equivalent } MW \times Q\text{-factor of the polymer}/41.3 \qquad (7.35)$$

The Q-factor approach does not give actual MW averages of the polymer only approximate and rough values, because the sizes of different kinds of polymers with the same MW in solution are not the same and depend on experimental conditions. MW averages obtained using the Q-factor approach are normally smaller than the absolute MW obtained by physical methods. Comparison of the PS equivalent MW of a polymer and the MW of the polymer obtained by physical methods gives an experimental Q-factor which is more practical for obtaining actual MW averages of the polymer. Examples of Q-factors experimentally obtained for several polymers are: polyethylene – 17.7, polypropylene – 26.4, PVC – 24.8, PMMA – 53.7. The sizes of different kinds of polymers do not bear a constant relationship to one another. When the Q-factor of PS is assumed to be 41.3 over the entire MW range, the Q-factor for PMMA varies with MW as follows: MW ($\times 10^{-4}$) of PMMA vs. Q-factor, 0.50 – 33.0; 1.0 – 43.4; 5.0 – 49.6; 10.0 – 53.7; 50.0 – 60.7; 100.0 – 62.0.

In the case of polymers having heteroatoms or a phenyl group in the chain, such as polycarbonates and polyesters, the extended chain length per monomer unit is not 2.52 Å. It must be calculated from the bond lengths and valence angles of C – O – C or a phenyl group. Q-factors for aliphatic polyesters calculated by comparison of MW measured by vapor-pressure osmometry and that by SEC using a PS calibration curve in THF at 37 °C are as follows: poly(ethylene adipate) 21.8 – 20.6; poly(ethylene subelate) 30.0 – 28.2; poly(ethylene sebacate) 40.8 – 31.4; poly(tetramethylene adipate) 18.8, poly(hexamethylene adipate) 14.3.

7.5 Universal Calibration

7.5.1 Intrinsic Viscosity

Intrinsic viscosity [η] is an experimental quantity derived from the measured viscosity of a polymer solution. Solution viscosity is usually measured in capillary viscometers. The ratio of the efflux time t required for a specified volume of a polymer solution to flow through a capillary tube to the corresponding efflux time t_0 for the solvent gives the relative viscosity

$$\eta_r = t/t_0 = \eta/\eta_0 \tag{7.36}$$

where η and η_0 are the viscosities of the volume solution and the solvent, respectively. The specific viscosity η_{sp} is

$$\eta_{sp} = (\eta - \eta_0)/\eta_0 = \eta_r - 1 \tag{7.37}$$

The reduced viscosity η_{red} and the inherent viscosity η_{inh} are

$$\eta_{red} = \eta_{sp}/c \tag{7.38}$$

and

$$\eta_{inh} = (\ln \eta_r)/c \tag{7.39}$$

where c is the concentration of the solution (g/ml).

The intrinsic viscosity is defined as

$$[\eta] = \lim_{c \to 0} \eta_{red} = \lim_{c \to 0} \eta_{inh} \qquad (7.40)$$

The intrinsic viscosity is independent of concentration because of extrapolation to $c = 0$ and is a function of the solvent and temperature. The intrinsic viscosity of a polymer is usually determined by measuring the relative viscosities for polymer solutions of varing concentration and extrapolating a plot of the reduced viscosity vs. concentration to zero concentration. This is time consuming, however, so that a single viscosity measurement at low sample concentrations is employed for convenience.

The Huggins equation for obtaining the intrinsic viscosity is

$$\eta_{sp}/c = [\eta] + k'[\eta]^2 c \qquad (7.41)$$

where k' is the Huggins constant for a series of polymers of different concentrations in a given solvent. If k' is known, or once k' is experimentally determined using Eq. (7.41), then the intrinsic viscosity of a polymer solution can be calculated using the following equation:

$$[\eta] = \frac{-1 + (1 + 4k'\eta_{sp})^{1/2}}{2k'c} \qquad (7.42)$$

Similarly, the Kraemer equation is:

$$\ln \eta_r/c = [\eta] + k''[\eta]^2 c \qquad (7.43)$$

where k'' is the Kraemer constant. Combining Eqs. (7.41) and (7.43) with the assumption that $k' - k'' = 0.5$ leads to the Solomon–Ciuta equation:

$$[\eta] = \{2(\eta_{sp} - \ln \eta_r)\}^{1/2}/c \qquad (7.44)$$

This equation can also be used to obtain the intrinsic viscosity by a single viscosity measurement. This method has been shown to give excellent agreement for dilute solutions (less than 0.5%) of polymers in the MW range $10^4 - 10^6$.

The value of the intrinsic viscosity for a polymer in a specific solvent is related to the MW of the polymer as

$$[\eta] = KM^a \qquad (7.45)$$

where K and a are constants determined from a double logarithmic plot of intrinsic viscosity and MW. This equation is called the Mark–Houwink equation. The Mark–Houwink constants K and a vary with polymer type, solvent, and temperature. For random coil polymers, the exponent a varies from 0.5 in a theta-solvent to a maximum of about 1.0. Typical values of K range between 5×10^{-3} and 5×10^{-2} (ml/g). Pairs of K and a values for a wide variety of polymers, solvent, temperature combinations are tabulated in Appendices I–III.

7.5.2 Universal Calibration Curve

A peak position calibration curve, which plots log MW against retention volume, is different when the polymer type is different. This is because the molecular size of different types of polymers in solution cannot simply be related to MW. Examples are shown in Fig. 7.4 for PS, PVC and comb-like PS. There are three curves for the three types of polymers.

Plots of $\log[\eta]M$ vs. retention volume for all polymers of different types including different conformations merge into a single plot and this curve, introduced by Benoit et al. [5], is called the universal calibration curve. An example of a calibration curve is also shown in Fig. 7.4 for PS, PVC, and comb-

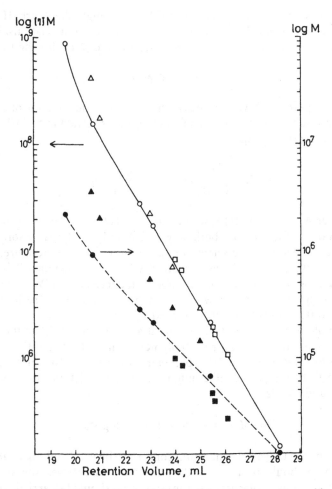

Fig. 7.4. Universal calibration curve ($\log[\eta]M$ vs. V_R) and peak position calibration curve ($\log M$ vs. V_R). Universal calibration curve: ○ PS, □ PVC, △ PS (comb); peak position calibration curve: ● PS, ■ PVC, ▲ PS (comb)

like PS. The solvent used is tetrahydrofuran (THF). All the plots for these three types of polymer almost fit on a single curve.

7.5.3 Transformation of the Universal Calibration Curve

The product of intrinsic viscosity and MW, $[\eta]M$, is proportional to the hydrodynamic volume and can be used as a universal parameter in SEC calibration. At any retention volume, the hydrodynamic volumes of two polymers A and B will be equal according to

$$[\eta]_A M_A = [\eta]_B M_B \tag{7.46}$$

The intrinsic viscosity is related to the MW through the Mark–Houwink equation, Eq. (7.45), and Eq. (7.46) can be converted into a relationship containing Mark–Houwink constants by substituting Eq. (7.45) for the two polymers.

$$K_A M_A^{a_A+1} = K_B M_B^{a_B+1} \tag{7.47}$$

The MW of polymer B which elutes at the same retention volume of polymer A in the same solvent and at the same temperature is related to the MW of polymer A by rearranging Eq. (7.47) to give

$$\log M_B = \frac{1}{1+a_B} \log \frac{K_A}{K_B} + \frac{1+a_A}{1+a_B} \log M_A \tag{7.48}$$

If polymer A is PS or PMMA, where narrow-MWD standards are available, and when values of K and a for both polymers A and B in the same solvent and at the same temperature are known or can be found in the literature, then a calibration curve of log MW vs. retention volume for polymer B by using Eq. (7.48) can be constructed from a calibration curve of log MW vs. retention volume for polymer A, or by knowing the retention volumes of a series of polymer A with different MWs together with Eq. (7.48).

An example is shown for PVC measured at 25 °C with THF as the mobile phase. The values of K and a for PS (polymer A) at 25 °C with THF are 1.23 × 10^{-2} and 0.705 and those for PVC (polymer B) are 1.59 × 10^{-2} and 0.753, respectively. Substituting these values into Eq. (7.48), the relationship between the MWs of PS and PVC is

$$\log M_{PVC} = -0.0636 + 0.973 \log M_{PS}$$

In order for the universal calibration concept to be valid and Eq. (7.48) to be effective, there should be no secondary interactions between the stationary phase and the sample molecules and the separation should be carried out solely by size exclusion. These secondary interactions (adsorption, ion exclusion, etc.) are often observed in aqueous SEC systems for ionic polymers or non-aqueous

SEC systems for polar polymers. Besides these problems, the selection of pairs of K and a values for a sample polymer and a standard polymer is also important. Many pairs of K and a values for PS can be found in Appendix I and selection of a pair of values may be confusing. Inaccurate or inappropriate pairs of K and a values are often found in the literature. The source and limitations on the use of K and a values should be noted when they are used in calculations. A preferable selection rule for several polymers is to obtain pairs of K and a values for the polymer under investigation and for PS or other calibration standards from the same source.

In a dilute solution of a polymer in a good solvent, the interaction between the solvent and the polymer is predominant as compared to a theta-solvent. In order to compensate for this interaction, the use of the value $[\eta] M / f(\varepsilon)$ instead of the product $[\eta] M$ is proposed, where

$$f(\varepsilon) = 1 - 2.63\varepsilon + 2.86\varepsilon^2 \tag{7.49}$$

$$\varepsilon = (2a - 1)/3$$

and a is the exponent in the Mark–Houwink equation. Eq. (7.48) can be converted to

$$\log M_B = \frac{1}{1 + a_B} \log \frac{K_A f(\varepsilon_B)}{K_B f(\varepsilon_A)} + \frac{1 + a_A}{1 + a_B} \log M_A \tag{7.50}$$

It becomes identical to Eq. (7.48) at the theta-point ($a = 0.5, \varepsilon = 0$) [6].

7.5.4 Direct Calculation of MW with Universal Calibration

Section 7.5.3 explained the approach to the construction of a calibration curve for a polymer of interest from averages using a peak position calibration curve for a polymer standard with Eq. (7.48). In this section, the direct calculation of MW averages through a universal calibration curve without using Eq. (7.48) is discussed.

The parameter J is defined as the product of the intrinsic viscosity and MW for the sake of simplicity where,

$$J_i = [\eta]_i M_i \tag{7.51}$$

The intrinsic viscosity, weight-average MW and number-average MW of the polymer concerned are defined as

$$[\eta]_i = K M_i^a \tag{7.52}$$

$$[\eta] = \sum W_i [\eta]_i \tag{7.53}$$

$$M_n = 1/\sum (W_i/M_i) \tag{7.54}$$

$$M_w = \sum W_i M_i \tag{7.55}$$

where W_i, $[\eta]_i$, and M_i are the weight fraction, intrinsic viscosity, and MW of the

ith point of the retention volume. Subtraction of M_i from Eqs. (7.51) and (7.52) and rearrangement yields:

$$[\eta]_i = J_i^{a/(1+a)} K^{1/(1+a)} \tag{7.56}$$

which, when substituted into Eq. (7.53), yields for the whole polymer

$$[\eta] = K^{1/(1+a)} \sum (W_i J_i^{a/(1+a)}) \tag{7.57}$$

Rearrangement of Eqs. (7.51) and (7.57) and substitution into Eqs. (7.54) and (7.55) give

$$M_n = K^{-1/(1+a)} / \sum (W_i / J_i^{1/(1+a)}) \tag{7.58}$$

and

$$M_w = K^{-1/(1+a)} \sum (W_i J_i^{1/(1+a)}) \tag{7.59}$$

J_i can be obtained from the universal calibration curve for PS, for example, and K and a are the values for the polymer under investigation.

7.5.5 Use of Secondary Standard(s) with Universal Calibration

The calibration of MW averages by universal calibration is usually dependent on the knowledge of the Mark–Houwink constants for both polymers: a polymer standard and the polymer of interest. When K and a for the polymer of interest are not known, and if there are two broad-MWD polymers (a secondary standard) having different intrinsic viscosities, then the construction of a calibration curve for the polymer or the calculation of MW averages is possible [6].

The intrinsic viscosities and SEC chromatograms of two polymers (a secondary standard) are measured and the ratio of the intrinsic viscosities is calculated. From Eq. (7.57)

$$[\eta]_1 / [\eta]_2 = \sum W_{1,i} J_i^{a/(1+a)} / \sum W_{2,i} J_i^{a/(1+a)} \tag{7.60}$$

Suffix 1 and 2 denote polymer 1 and polymer 2 of interest and J_i can be obtained from the universal calibration curve of PS (a primary standard). Assume the exponent a of the Mark–Houwink equation to initially be in the range 0.5 to 1.0, calculate Eq. (7.60), and repeat the assumption of the value a by a trial-and-error search until both sides of Eq. (7.60) are equal. The value of a thus obtained is substituted into Eq. (7.57) and the value of K is calculated. The next step is to follow the instructions in Sect. 7.5.3 or 7.5.4.

When one broad-MWD sample with known M_w is available, Eq. (7.61) is used. The intrinsic viscosity should be measured first, then Eq. (7.61) iterated by assuming an exponent a and repeating the process until both sides are equal.

$$[\eta] M_w = \sum (W_i J_i^{a/(1+a)}) \sum (W_i J_i^{1/(1+a)}) \tag{7.61}$$

Substitute the final value of a into Eq. (7.59), and the coefficient K can be obtained. Similarly, when one sample with known M_n is available, then the following equation can be used:

$$[\eta] M_n = \sum (W_i J_i^{a/(1+a)}) \sum (W_i J_i^{1/(1+a)}) \tag{7.62}$$

When two broad-MWD polymers have known M_w or M_n values, then one of the following three equations can be applied in the same manner:

$$M_{w,1}/M_{w,2} = \sum (W_{1,i} J_i^{1/(1+a)})/\sum (W_{2,i} J_i^{1/(1+a)}) \tag{7.63}$$

$$M_{n,1}/M_{n,2} = \sum (W_{2,i}/J_i^{1/(1+a)})/\sum (W_{1,i}/J_i^{1/(1+a)}) \tag{7.64}$$

$$M_{w,1}/M_{n,2} = \sum (W_i J_i^{1/(1+a)}) \sum (W_i/J_i^{1/(1+a)}) \tag{7.65}$$

It should be pointed out that the approaches outlined in Sects. 7.5.3 and 7.5.4 depend on the validity of the Mark–Houwink constants for both the polymer standard and the polymer of interest, but the approach in Sect. 7.5.5 does not depend on these same constants for the primary polymer standard. However, when a pair of Mark–Houwink constants for a secondary standard obtained by this approach is used to construct a calibration curve or to calculate MW averages, the same pair of Mark–Houwink constants for the primary standard used to find the constants for the secondary standard should be used together.

References

1. BALKE ST, HAMIELEC AE, LECLAIR BP, PEARCE SL (1969) Ind Eng Chem Res Devel 8:54
2. MCCRACKIN FL (1977) J Appl Polym Sci 21:191
3. SWARTZ TD, BLY DD, EDWARDS AS (1972) J Appl Polym Sci 16:3353
4. MORI S (1981) Anal Chem 53:1815
5. GRUBISIC Z, REMPP P, BENOIT H (1967) J Polym Sci Polym Lett Ed 5:753
6. COLL H, GILDING DK (1970) J Polym Sci A-2, 8:89
7. WEISS AR, COHN-GINSBERG E (1969) J Polym Sci B7:379

8 Molecular-Weight-Sensitive Detectors

8.1 Light Scattering

8.1.1 Theory

When radiation passes through a transparent medium, the species in the medium scatter a fraction of the beam in all directions. The wavelength of the scattered radiation is almost the same as that of the incident radiation (Rayleigh scattering). The intensity of the scattered radiation is related to the molecular weight, the molecular size, the number, and the shape of the species in the medium as well as the scattering angle.

In light-scattering measurements, the reduced scattering intensity at a scattering angle θ, R_θ, is measured which is simply denoted as the Rayleigh ratio

$$R_\theta = (I_\theta r^2)/(I_o V) \tag{8.1}$$

where I_o is the intensity of the incident radiation, I_θ the intensity of the scattered radiation at a scattering angle θ, r the distance from the scattering center to the observer (detector), and V is the scattering volume of the sample solution. The excess Rayleigh ratio, $R(\theta)$, is defined as

$$R(\theta) = R_\theta - R_o \tag{8.2}$$

where R_o is the Rayleigh ratio of the solvent.

For a dilute solution of a monodisperse polymer with molecular weight M, $R(\theta)$ is expressed as

$$\frac{K^* c}{R(\theta)} = \frac{1}{MP(\theta)} + 2A_2 c + 3A_3 c^2 + \cdots \tag{8.3}$$

where C is the concentration of the polymer in the solution in g/ml, and A_2 and A_3 are the second and third virial coefficients, respectively. $P(\theta)$, which is defined as the ratio R_θ/R_o and is termed the particle scattering constant, is given by

$$P(\theta) = 1 - q^2 \langle S^2 \rangle /3 \tag{8.4}$$

where $\langle S^2 \rangle$ is the squared radius of gyration of the polymer. K^* is the optical constant, and for unpolarized incident light, it is given by

$$K^* = (2\pi^2 n_0^2 (dn/dc)^2)(1 + \cos^2\theta)/(N_A \lambda_o^4) \tag{8.5}$$

where n_0 is the refractive index of the solvent, dn/dc the specific refractive index increment, N_A is Avogadro's number, and λ_o the wavelength in vacuo of the incident light. For vertically polarized incident light

$$K^* = (4\pi^2 n_0^2 (dn/dc)^2)/(N_A \lambda_0^4) \tag{8.5'}$$

The term q is defined as

$$q = (4\pi n_o/\lambda_o)\sin(\theta/2) \tag{8.6}$$

In the derivation of Eq. (8.3), the intensity of the scattered light is proportional to the mass of the scattering particles. If the solute is polydisperse, the heavier molecules contribute more to the scattering than the lighter ones. The total scattering at zero concentration is

$$R(\theta) = K^* \sum c_i M_i = K^* c M_w \tag{8.7}$$

in which the weight-average molecular weight M_w is defined according to any of the relationships in the following equation:

$$M_w = \sum(c_i M_i)/\sum c_i = \sum(c_i M_i)/c$$
$$= \sum(N_i M_i^2)/\sum(N_i M_i) \tag{8.8}$$

Similarly, the squared radius of gyration can be shown to be a z-average quantity, $\langle S^2 \rangle_z$. For a solution of a polymer with a polydispersity, Eq. (8.3) is modified as

$$\frac{K^* c}{R(\theta)} = \frac{1}{M_w}[1 + (16\pi^2 n_0^2/3\lambda_0^2)\langle S^2 \rangle_z \sin^2(\theta/2)] + 2A_2 c + 3A_3 c^2 + \dots \tag{8.9}$$

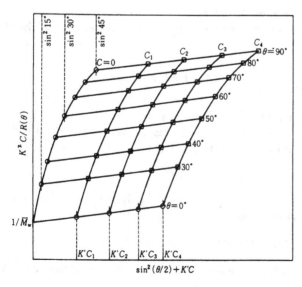

Fig. 8.1. Example of a Zimm plot. □: data point; ○: extrapolated points to zero concentration at constant angle and to scattering angle of 0° at fixed concentration

In order to estimate the values of M_w, $\langle S^2 \rangle_z$, and A_2 by light-scattering measurements, values of $K^*c/R(\theta)$ at various scattering angles and concentrations at constant temperature are plotted against $\sin^2(\theta/2) + K'c$, where K' is an arbitrary constant chosen to produce a reasonable spread of data points. A value of $K' = 1/c_{max}$, where c_{max} is the maximum concentration used, is usually employed. This plot is called the Zimm plot and an example is shown in Fig. 8.1. Extrapolation to zero concentration at constant angle yields the values of $K^*c/R(\theta = x)$ at $c = 0$ and that to a scattering angle $0°$ at fixed concentration gives the values of $K^*c/R(\theta = 0)$. The value of M_w is obtained as the inverse of the value of the intercept of a line of $K^*c/R(\theta = x)$ at $c = 0$ vs. $\sin^2(\theta/2)$ and that of $Kc/R(\theta = 0)$ vs. $K'c$. The initial slope of the plot of $K^*c/R(\theta = x)$ at $c = 0$ vs. $\sin^2(\theta/2)$ at a scattering angle of almost $0°$ gives the value of $\langle S^2 \rangle_z$ by dividing the slope by $(16\pi^2 n_0^2/3\lambda_0^2 M_w)$. A plot of $K^*c/R(\theta = 0)$ vs. $K'c$ at almost zero concentration, divided by 2, gives the value of A_2.

8.1.2 Instrumentation

From the late 1950s to the early 1960s, light-scattering (LS) instruments became commercially available. LS photometers extensively used as SEC detectors appeared in the mid-1970s and presently there are five laser-based LS photometers that allow the measurement of scattering intensities at different angles on the market: low-angle laser LS (LALLS), multi-angle laser LS (MALLS), triple-angle laser LS (TALLS), dual-angle laser LS (DALLS), and right-angle laser LS (RALLS) photometers.

The first commercially available LALLS photometer was the Chromatix KMX-6 instrument. In this instrument, the incident beam is vertically polarized and the source is a 2-mW helium neon (He-Ne) laser operating at $\lambda_0 = 632.8$ nm. The incoming light is focused by a condensing lens onto a 100-μm diameter target in the sample cell which consists of two thick fused-silica windows separated by a spacer with a hole in the center. The light scattered by the sample through an angle θ of less than $10°$ is collected onto a field stop, and is detected by a photomultiplier. In LALLS measurements, the scattered angle is assumed to be almost zero, so that Eq. (8.9) can be simplified as

$$K^*c/R(\theta) = (1/M_w) + 2A_2c \qquad (8.10)$$

The scattering angle is defined by the solid annulus which can adjust the scattering angles of $2-3°$, $3-4°$, $4.5-5.5°$, $6-7°$, and $3-7°$, respectively.

The DAWN Model F LS photometer (MALLS) (Wyatt Technology, USA) is capable of measuring intensities at up to 18 angles simultaneously, thus allowing the measurement of the radius of gyration as well as M_w. The flow cell consists of a 2-mm channel drilled through a cylindrical cell body. The 5-mW linearly polarized He-Ne laser is directed along the length of the cell. Eighteen photodiode detectors are placed around the cell at fixed detector angles, θ', such that they are equally spaced in cotangent θ'. Therefore, the fixed detector angles are between 26.56 and 144.46°. The real scattering angles, θ, can be calculated from the refractive index of the solvent. The choice of solvent determines the smallest usable detection angle. The user is guided by the software to select a set of detectors, the lowest of which lies at or above a minimum detection angle. The use of 18 or fewer angles is also possible. The light source is a 5-mW He-Ne laser at 632.8 nm. The analyzing software AURORA is used to compute $M_w, \langle S^2 \rangle$, and A_2 by Zimm plots.

The TALLS photometer (mini DAWN) is a version of the Dawn model F and has only three photodiode detectors placed at 45, 90, and 135°. This instrument uses a 20-mW semiconductor laser at a wavelength of 690 nm. The DALLS photometer (Precision Detectors, USA) measures the intensities of the scattered beam at 15 and 90°. The RALLS photometer (Viscotek Corp., USA) uses a laser diode (670 nm) as the light source and measures the intensity of the scattered beam at 90°.

8.1.3 Methodology

When combined with SEC, the LS intensity is measured only at a single concentration for each MW fraction eluting from the column. A concentration detector, such as a refractive index (RI) detector, is connected in series or in parallel after the LS instrument. In order to calculate the MW, the second virial coefficient A_2 must be known beforehand, or can be neglected if the concentration of the polymer solution injected is diluted considerably after passing through the column. The radius of gyration can only be measured for molecules greater than about 10 nm in diameter; below this size it is extremely difficult to measure the variation in scattered intensity with angles.

Optically clean solutions, especially aqueous solutions, must be carefully prepared for LS measurements. Filtration through a membrane filter of 0.2 μm (and 0.1 μm) is used to remove dust particles.

When LALLS is used as the LS detector, LALLS and concentration chromatograms are obtained. Because the scattering angle for LALLS is assumed to be nearly zero, the Rayleigh ratio at each increment i and the excess Rayleigh ratio is measured as the peak intensity from the baseline. The MW at each increment i can be calculated by

$$M_i = [(K^* c_i / R(\theta)_i) - 2A_2 c_i]^{-1} \tag{8.11}$$

The value of c_i can be obtained from the response of the concentration detector, if the relationship between the detector response and the concentration is known. Alternatively, the value of c_i can be obtained from the following equation:

$$c_i = m X_i / (\Delta V \sum X_i) \tag{8.12}$$

where m is the sample mass injected, X_i the response of the concentration detector, and ΔV the retention volume increment between adjacent data points. Eq. (8.11) can be rewritten as

$$M_w = \sum (M_i c_i) / \sum c_i = (\sum R(\theta)_i / K^*) / \sum c_i \tag{8.13}$$

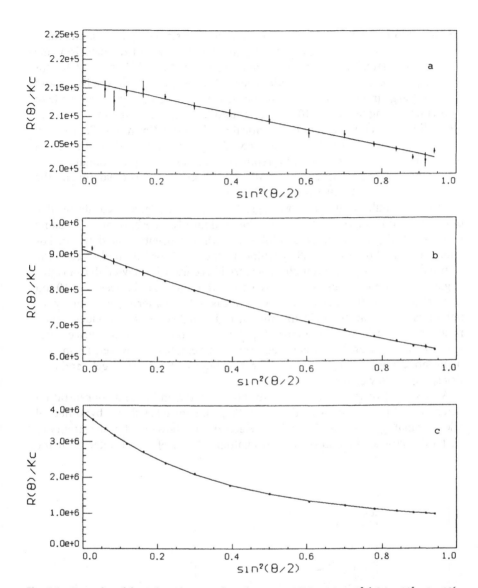

Fig. 8.2. Examples of the Debye plot. Sample: polystyrene; MW: a 2.2 × 10^5, b 9.1 × 10^5, c 4 × 10^6

When MALLS is used as the LS detector with a concentration detector, the analyzing software ASTRA calculates M_w and $\langle S^2 \rangle_z$. The ASTRA software uses the Debye method for calculating MWs and radii. The Debye method uses a fit of $R(\theta)/K*c$ vs. $\sin^2(\theta/2)$ for each slice, while the Zimm method uses a fit of $K*c/R(\theta)$ vs. $\sin^2(\theta/2) + K'c$. The difference is in the value of the y-axis, because one plot is the reciprocal of the other. However, the Debye method may be adequate for polymers having MW over 10^6. Examples of use of the Debye plot are shown in Fig. 8.2. For molecules with molecular sizes less than 10 nm, there is no measurable scattering asymmetry and the Debye plot displays isotropy throughout scattering angles. This size corresponds to a MW of about 10^5 for flexible polymers. The plots for mid-size molecules (10–30 nm) are linear and those for large molecules over 30 nm curved. Figure 8.2a shows an example of a polymer having MW 2.2×10^5 and a radius of gyration near 16 nm. The Debye plot is linear and gives an almost identical result to the Zimm plot. Figure 8.2b shows an example of a polymer having MW 9.1×10^5 and a radius of gyration 43 nm. In this case, the Debye plot requires a second-order polynomial. Figure 8.2c is the case of a polymer having MW 4×10^6 and a radius of 96 nm and the Debye plot requires a fourth-order polynomial.

Although multi-angle and low-angle LS instruments give accurate results, other types of LS instruments can also be used for MW measurements by using them as LS detectors for SEC combined with a concentration detector. For triple-angle and dual-angle LS instruments, the Debye or Zimm plots can only be fitted to a linear line, although a decrease in accuracy is unavoidable. A right-angle LS instrument is useful for polymers with molecular size less than 10 nm (MW about 100000), because there is no measurable scattering asymmetry for polymers lower than 10 nm in size and the right-angle intensity provides a good measurement of MW. For higher MW polymers, a combination of a viscometer and a concentration detector is required (see Sect. 8.3). However, LS instruments are generally very useful for detecting high-MW materials, gels, and particles, even at low concentration.

An LS detector is more sensitive to the high-MW end than a concentration detector, and less sensitive to the low-MW end. Figure 8.3 shows hypothetical SEC chromatograms from LS and RI detectors and shows the low sensitivity of an LS detector at the end of the chromatogram and of an RI detector at the

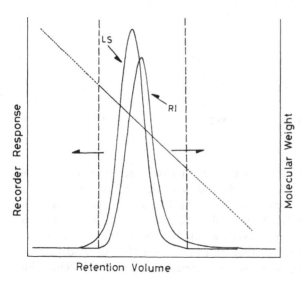

Fig. 8.3. Hypothetical SEC chromatograms from light-scattering (LS) and refractive index (RI) detectors and the relationship of molecular weight and retention volume

onset of the chromatogram. Therefore, a reliable MW estimation at both ends of the chromatogram is impossible. The areas within the dashed lines indicate the reliable regions of the chromatograms for data analysis and shows that the determination of MW at the ends of a chromatogram results in high errors. Because the relationship of MW vs. retention volume (solid line) obtained at the center of a chromatogram is reliable, linear extrapolation to both ends (dotted lines) can provide a good approximation of the relationship. Thus, the unobserved low-MW portion of the LS response curve may be accounted for by considering the concentration data from an RI detector and the estimation for number-average MW, M_n, is improved. However, the high-MW components cannot be accounted for and thus M_w will be underestimated.

Alternatively, the area can be used for the calculation of M_w instead of a point-by-point summation of calculated MW. This method has been shown to give greater precision than the summation of individual values at each retention volume and can be used for samples containing a high-MW fraction that is detected only by an LS detector. The value of M_w can be given by

$$M_w = \Delta V \sum R(\theta = 0)_i / (K^* m) \tag{8.14}$$

where ΔV is the volume increment between data points and m is the sample mass injected [1]. Alternatively,

$$M_w = k_c \times \text{total peak area of the LS signal/}$$

$$K^* k_{LS} \times \text{total peak area of the RI signal} \tag{8.15}$$

where k_c is the proportionality constant between the signal of the RI detector and the concentration, and k_{LS} is a similar constant for the LS detector signal [2].

8.1.4 Problems of Refractive Index

In order to determine MW by LS measurement, the refractive index n_o of the solvent and the specific refractive index increment, dn/dc, for the solution must be known. As can be seen in Eq. (8.5), the accuracy of the LS measurement depends on dn/dc of the sample in the solvent. An error in dn/dc of 2% leads to a corresponding error in M_w of 4%. The solvent refractive index is measured with a conventional refractometer or values found in the literature. The dn/dc value can be measured either by a differential refractometer or an interferometer. Measurements should be made at the same temperature as the LS measurement and at the same wavelength.

If the polymer refractive index, n_p, and the partial specific volume of the polymer in the solvent, v_p, are known, then dn/dc can be estimated by the Gladstone–Dale rule [3]

$$dn/dc = v_p(n_p - n_o) \tag{8.16}$$

The value of v_p is nearly unity. Values of dn/dc have a linear dependence on the solvent refractive index so that if the required value is not available in the solvent to be used, it can also be determined from Eq. (8.17) from other solvent systems

$$(dn/dc)_b = (dn/dc)_a - v_p(n_b - n_a) \qquad (8.17)$$

where suffixes a and b denote solvent a and solvent b, respectively.

The wavelength dependence of dn/dc is not negligible. The value at the desired wavelength can be obtained by extrapolation of a plot of dn/dc against the inverse of the wavelength squared, using the following relationship:

$$dn/dc = s + t/\lambda^2 \qquad (8.18)$$

where s and t are the intercept and slope, respectively. The value for dn/dc also varies with MW as

$$dn/dc = d + e/M \qquad (8.19)$$

where d and e are constants and d corresponds to the specific refractive index increment at infinite MW. The value of dn/dc increases with increasing MW and reaches an asymptotic limit for MWs greater than approximately 2×10^4. The experimental values for dn/dc for several polymers are tabulated in Appendix IV.

The dn/dc values for copolymers vary with the copolymer compositions. The average value of dn/dc for a copolymer is written as

$$(dn/dc)_C = W_A(dn/dc)_A + W_B(dn/dc)_B \qquad (8.20)$$

where W_A and W_B are weight fractions of A and B units in the copolymer and C denotes the copolymer. If the composition of the copolymer is homogenous over the entire range of MW and the refractive index of homopolymer A is similar to homopolymer B, and is far from that of the solvent, the M_w measured by LS can be assumed to be a true one. In most cases, the composition is heterogeneous and only an apparent M_w is measured, which depends on the solvent refractive index. In order to determine the true M_w, LS measurements must be carried out in at least three solvents with different refractive indices.

The difference between the true M_w and the apparent M_w increases with an increase in the width of the chemical composition distribution of the copolymer and an increase in the difference between the refractive index increments of the corresponding homopolymers [3]. The MW averages and the MW distribution calculated from the data obtained in SEC measurement with combined LS and concentration detectors are very good approximations of the correct distribution, unless the differences of the refractive index increments of the corresponding homopolymers are extremely high, which seldom happens in practice [4].

8.2 Viscometry

8.2.1 Theory

MW determination by viscosity is based on the fact that the viscosity of a polymer solution, η, is generally larger than that of the solvent, η_o, and depends on the MW of the polymer in the solution. Viscosity, η, is expressed in the region of low solution concentration c by:

$$\eta = \eta_o(1 + [\eta]c + K_\eta c^2) \qquad (8.21)$$

where $[\eta]$ is the intrinsic viscosity and K_η is a concentration-independent parameter. The relative viscosity η_r and the specific viscosity η_{sp} are defined in Eqs. (7.36) and (7.37). The reduced viscosity, η_{red}, the inherent viscosity, η_{inh}, and the intrinsic viscosity are defined in Eqs. (7.38), (7.39) and (7.40), respectively.

The viscosity of polymer solutions is usually measured by capillary or rotational viscometers. Typical capillary-type viscometers in common use are Ostwald, Ubbelohde, and Cannon–Fenske types. However, these viscometers cannot be employed as SEC detectors. Automatic capillary viscometers that are used as SEC detectors have been introduced recently.

When a polymer solution having a viscosity η flows through a capillary of length l with a radius r under laminar flow, the pressure drop ΔP between the ends of the capillary is given by the Hagen–Poiseuille law as

$$\Delta P = 8lQ\eta/(\pi r^4) \qquad (8.22)$$

where Q is the flow rate of the fluid. When the pressure drop of the mobile phase across the capillary is ΔP_o, the difference between the pressure drop of ΔP and ΔP_o is expressed as

$$\Delta\Delta P = \Delta P - \Delta P_o = (8lQ/\pi r^4)(\eta - \eta_o) \qquad (8.23)$$

where η_o is the viscosity of the mobile phase.

For very dilute polymer concentrations, such as those in SEC, the intrinsic viscosity of the polymer sample is defined as

$$[\eta] = \lim_{c \to 0} (1/c)(\eta - \eta_o)/\eta_o = (1/c)(\Delta P - \Delta P_o)/\Delta P_o$$

$$= (1/c)\Delta\Delta P/\Delta P_o \qquad (8.24)$$

where c is the sample concentration and is obtained from the response of the concentration detector.

8.2.2 Instrumentation

Several types of capillary viscometers (CV), which measure the pressure drop across a capillary or the differential pressure across two capillaries, are now commercially available. These viscometers are used extensively as a MW-sensitive detectors combined with a concentration detector for SEC.

A schematic diagram of a CV is shown in Fig. 8.4. A differential pressure transducer is connected to the inlet and outlet of the capillary and is used to monitor the pressure drop of the fluid flowing through the capillary. The single-capillary design commercially available uses a capillary of 15-cm length and 0.35-mm i.d. The capillary is enclosed in a dual-wall container to isolate it from any temperature fluctuations. From tees at either end of the capillary, stainless-steel tubing is connected to each side of a variable reluctance differential pressure transducer, of which the full-scale range is 5 KPa. At a flow rate of 1 ml/min with tetrahydrofuran at 35 °C, the pressure drop across the capillary due to pure solvent is about 50% of full scale, i.e. about 2.5 KPa. During elution of a polymer under typical chromatographic conditions, this signal increases by a maximum of about 1% of the background signal. Computational procedures, including Fourier-filtering of the raw viscometer data, are required to selectively remove periodic noise at the frequencies of operation of the piston pump system and high frequency noise from other sources.

The major problems encountered in the operation of the single-capillary viscometer are unavoidable flow-rate and temperature fluctuations in addition to pump pulsation. To overcome the problem of flow-rate dependency of the viscometer response, two sets of capillary and differential pressure transducer

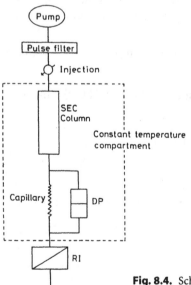

Fig. 8.4. Schematic diagram of a capillary viscometer with one capillary. *DP* Differential pressure transducer

assemblies are utilized [5]. This type of CV is commercially available. The capillaries are connected in series. At the time the sample solution from the SEC columns passes through the analytical capillary (the first capillary), the mobile phase solvent continuously flows through the reference capillary (the second capillary). A delay volume is added between the analytical and the reference capillaries. The function of the delay volume is to prevent the sample solution reaching the reference capillary during the time that the sample viscosity is being monitored in the analytical capillary. The delay volume element is nothing more than a large i.d. coiled tubing. The differential pressure signals from the two capillary-transducer systems are fed to a differential logarithmic amplifier to give $\log \eta_r$. Since the flow rates in the two capillaries connected in series have to be the same, flow rate effects are cancelled out to give flow-rate independent signal.

The third commercially available CV uses a parallel bridge design of four capillaries and measures the differential pressure across two capillaries, one for the sample solution and the other for the mobile phase, and can monitor the differential pressure directly [6].

The laboratory-made capillary viscometers reported in the literature contain one capillary [7, 8], two capillaries [9], or four capillaries [10] and are attractive due to their relatively simple design, ease of data reduction, and low cost, as compared with commercial viscometers.

8.2.3 Methodology

Measurement of the specific viscosity or relative viscosity requires that both the solution and the solvent viscosity be measured at the same flow rate. This can be achieved by measuring the solvent viscosity as the baseline before and after the polymer peak elutes, and by measuring the solution viscosity or by measuring the solution viscosity using an analytical capillary and the solvent viscosity using a reference capillary.

The intrinsic viscosity of a polymer fraction eluted at each retention volume increment i is obtained as

$$[\eta]_i = (1/c_i)\, \Delta\Delta P_i / \Delta P_o \qquad (8.25)$$

or

$$[\eta]_i = (1/c_i)\, \ln(\Delta P_i / \Delta P_o) \qquad (8.26)$$

When the concentration chromatogram of a sample is divided into intervals y, then c_i is calculated as

$$c_i = m h_i / (y \sum h_i) \qquad (8.27)$$

where m is the mass of the sample injected into the SEC column system and h_i is the height of the concentration chromatogram at increment i. The units of y and m will be dl or ml and g, respectively.

The intrinsic viscosity of the whole polymer can be calculated by using either of the following equations:

$$[\eta] = (y/m\Delta P_o) \sum \Delta\Delta P_i \qquad (8.28)$$

or

$$[\eta] = (y/m) \sum \ln(\Delta P_i/\Delta P_o) \qquad (8.29)$$

The product of the intrinsic viscosity and the MW of a polymer is proportional to the hydrodynamic volume V_h, and, in practice, it is expressed simply as

$$V_h = [\eta] M \qquad (8.30)$$

A universal calibration curve of polymer hydrodynamic volume against retention volume, which is valid for different types of polymers as well as for copolymers and branched polymers, can be constructed. This is achieved by using narrow-MW distribution standards with known MW and known intrinsic viscosities (or Mark–Houwink parameters) such as polystyrene standards. A calibration curve for a polymer of interest is reconstructed from the universal calibration curve using the following relationship:

$$M_i = V_{h,i}/[\eta]_i \qquad (8.31)$$

where M_i and $[\eta]_i$ are the MW and the intrinsic viscosity of the polymer at increment i and $V_{h,i}$ is the hydrodynamic volume obtained from the universal calibration curve already established for the SEC system prior to the sample analysis. MW averages and MW distributions of the polymer can be calculated using the reconstructed calibration curve. Alternatively, they can be calculated using the following equations:

$$M_n = \sum c_i / \sum (c_i/(V_h/[\eta]))_i \qquad (8.32)$$

and

$$M_w = \sum c_i(V_h/[\eta])_i / \sum c_i \qquad (8.33)$$

Alternatively, Eq. (8.32) can be rearranged using Eqs. (7.38), (7.39) and (7.40) as

$$M_n = \sum c_i / \sum (\eta_{sp}/V_h)_i \qquad (8.34)$$

or

$$M_n = \sum c_i / \sum (\ln \eta_r/V_h)_i \qquad (8.35)$$

Because the value of $\sum c_i$ is m/y from Eq. (8.32), Eqs. (8.39) and (8.35) can be expressed as

$$M_n = m/(y \sum (\eta_{sp}/V_h)_i) \qquad (8.36)$$

or

$$M_n = m/(y \sum (\ln \eta_r/V_h)_i) \qquad (8.37)$$

where y is the retention volume increment between data points. It is of interest to note that the value of M_n can be obtained without a concentration detector using Eq. (8.36) or (8.37).

8.3 Combination with a Concentration Detector

8.3.1 Methodology

The use of MW-sensitive detectors, such as an LS detector or a CV detector in conjunction with an RI detector, facilitates the direct calculation of MW averages of polymer samples without preconstruction of a calibration curve. There are three combinations for the use of MW-sensitive detector(s) and a concentration detector: (1) LS-RI, (2) CV-RI, and (3) LS-CV-RI. In combinations 1 and 2, the RI detector is placed in series after the LS or the CV detector. This is because the cell of an RI detector, which is normally used as a concentration detector for SEC, is fragile. An ultraviolet (UV) detector, however, can be connected between the SEC columns and the LS or the CV detector. Parallel connection of the LS or the CV detector and an RI detector is also possible.

In the case of combination 3, a series connection in the order of LS, CV, and RI detector is probable. Because of the large dead volume in the RI detector, it should not be placed before the CV detector. In the parallel configuration, the LS and the CV detectors are placed in parallel and the eluate from the SEC columns is split approximately equally. The RI detector is placed in series after the LS detector, or the CV and RI detectors are connected in parallel after the LS detector. The ratio of the volume of flow between the CV and the RI lines should be approximately 50:50. The eluate from the SEC columns is split approximately equally to the LS and the CV detector by regulating the back pressure on the CV branch. Series configurations provide greater control over flow rate fluctuations because of back-pressure variations among detectors. Parallel configurations avoid additional peak broadening caused by the eluting sample passing through a number of detector cells. However, band broadening and peak distortion are dependent on the detector cell design and the configurations of the connecting tubing and T-junctions used to split the flow.

When two or three detectors are used together, accurate estimation of the interdetector volume of liquid in the connecting tubing between the two detectors is very important. There are a number of approaches that can be used to determine the interdetector volume and the obvious procedure is to calculate the geometric offset volume from the connection volume between detectors. However, these calculated values are not accurate. The most commonly used approach for determining interdetector volume is to measure the peak maxima (or breakthrough volumes) difference of a narrow-MW distribution standard [11].

It should be kept in mind that these three detectors differ in sensitivity. The response of the LS detector is proportional to cM (c = concentration, M = molecular weight), while the CV detector responds according to $cM^{0.5}$ to about $cM^{0.8}$, depending on the given solvent [12]. The RI detector, however, scales with c. Both the LS and CV detectors are sensitive to high-MW species and insensitive to small molecules. When the largest molecules appear in the effluent, the LS and the CV detectors respond strongly, even though the concentration of the molecules is very small, whereas the RI detector indicates zero concentration.

In the low-MW region, the RI chromatogram indicates finite concentration while the LS and CV detector show zero response.

This mismatch of sensitivities may result in inaccurate MW averages. The overall calculation of the averages from integration of the LS and CV chromatograms without using the RI signal for local concentration should improve the problem of sensitivity mismatch.

8.3.2 Selected Applications

Light scattering has been widely used to study branching. The MALLS instrument can measure molecular sizes larger than 10 nm and therefore can give information about the branching of polymers. The branching factor, g, under theta-conditions for a given number and type (tri- or tetrafunctional) of branching points is defined as

$$g = (R_{g,\text{br}}^2 / R_{g,\text{l}}^2)_M \tag{8.38}$$

where R_g^2 is the mean square radius of gyration of the branched and linear polymers of the same MW. The subscripts l and br refer to linear and branched polymers, respectively. For different branching architecture, g can be related to the number of branches per molecule. For polymers having the same MW, branched polymers have smaller molecular sizes than linear ones. In other words, branched polymers having the same molecular size have larger MWs than linear ones, and thus g is less than unity.

For a branched and a linear polymer with the same MW, the ratio of intrinsic viscosities is expressed as

$$g^b = ([\eta]_{\text{br}} / [\eta]_{\text{l}})_M \tag{8.39}$$

The value of g^b for a linear polymer is unity and that for a branched polymer decreases as the number of branched points per molecule increases. Typical values for b range from 0.5 to 1.5. Eq. (8.39) can be transformed to

$$g^b = (M_{\text{l},i} / M_{\text{br},i})_{V_i}^{a+1} \tag{8.40}$$

where a is the Mark–Houwink exponent. The subscript V_i indicates that the MWs of both the linear and branched polymers are to be taken at the same retention volume increments. This approach is useful for estimating g^b if only an LS detector is employed with an RI detector. See Chap. 10 for more detailed information on branching.

The use of a CV detector in series with a RALLS detector allows a correction of the 90° LS data to zero angle. Advantages of the use of RALLS are simplified instrumentation, application of an HPLC fluorometer, higher signal-to-noise ratio compared to low-angle measurements, and less interference from particulates, although the data are less accurate. With the use of the Flory–Fox equation and intrinsic viscosity measurements, the following iterative approach can be used [13]:

1. Assume $P(\theta = 90) = 1$ and $A_2 = 0$.
2. Estimate MW from $M_{\text{est}} = R(\theta = 90)/Kc$.

3. Estimate R_g and $P(\theta = 90)$ from

$$R_{g,\text{est}} = (1/\sqrt{6})([\eta] M_{\text{est}}/\phi)^{1/3}$$

where $\phi = 2.55 \times 10^{23}(1 - 2.63\,\varepsilon + 2.86\,\varepsilon^2)$ and $\varepsilon = (2a - 1)/3$. Units of $[\eta]$ are ml/g. $P(\theta = 90)$ is estimated from

$$P(\theta = 90) = (2/X^2)(e^{-X} + X - 1)$$

where $X^{1/2} = (4\pi n_o/\lambda_o) R_{g,\text{est}} \sin\theta$.

4. Estimate MW again from the new value of $P(\theta = 90)$

$$M'_{\text{est}} = M_{\text{est}}/P(\theta = 90).$$

5. Back to 3 and repeat the iteration using M'_{est} until convergence is achieved.

The value $P(\theta = 90)$ is close to unity for globular proteins up to MW 10^6, random chain linear polymers up to 2×10^5, and for branched polymers over 2×10^5.

References

1. JENG L, BALKE ST, MOUREY TH, WHEELER L, REMEO P (1993) J Appl Polym Sci 49:1359
2. DEGROOT AW, HAMRE WJ (1993) J Chromatogr 648:33
3. KRATOCHVIL P (1987) Classical Light Scattering from Polymer Solutions. Elsevier, New York
4. KRATOCHVIL P (1995) The 8th International Symposium on Polymer Analysis and Characterization, Florida, USA, L14
5. YAU WW, ABBOT SD, SMITH GA, KEATING MY (1987) In: PROVDER T (ed) Detection and Data Analysis in Size Exclusion chromatography, ACS Symposium Series 352, ACS, Washington, DC, p. 80
6. HANEY MA (1985) J Appl Polym Sci 30:3023
7. LETOT L, LESEC J, QUIVORON C (1980) J Liq Chromatogr 3:427
8. MALIHI FB, KUO C, KOEHLER ME, PROVDER T, KAH AF (1984) In: PROVDER T (ed) Size Exclusion Chromatography, ACS Symposium Series 245, ACS, Washington, DC. p. 281
9. MORI S, HOUSAKI T (1996) Int J Polym Anal & Charact 2:335
10. MORI S (1993) J Chromatogr 637:129
11. MOUREY TH, MILLER SM (1990) J Liq Chromatogr 13:693
12. PANG S, RUDIN A (1992) Polymer 33:1949
13. HANEY MA, JACKSON C, YAU WW (1992) International GPC Symposium, Florida

9 Synthetic Polymers

9.1 Separation of Oligomers

9.1.1 Refractive Index and MW

A differential refractive index detector (RI refractometer) and an ultraviolet (UV) absorption detector are the most commonly used detectors for SEC. Molecular weight (MW) averages of a polymer are calculated by SEC on the premise that the magnitude of the ordinate (the response of the detector) on the chromatogram is proportional to the amount of the polymer which appears at each slice. The relationship between the refractive index n of a species and its MW is:

$$\frac{dn}{dc} = A + \frac{B}{M} \qquad (9.1)$$

where c is the concentration of the species, M is the MW of the species, and A and B are constants. Although the refractive index of a polymer with MW above 10000 is assumed to be independent of the MW, an increase in the refractive index for polystyrene (PS) with MW over 10000 has been reported [1]. The value of dn/dc of PS having MW 1800000 increases 9% to that of PS having MW 2000 in toluene and 4% in methyl ethyl ketone (MEK). The refractive index of toluene (1.496) is closer to that of PS (1.55–1.60) than that of MEK (1.3814). A solute and solvent pair having a smaller difference in refractive index has a larger MW dependency.

The MW dependence of the refractive index is more remarkable for oligomers. Figures 9.1 and 9.2 represent plots of refractive index against MW for oligo- (OEG) and poly(ethylene glycols) (PEG). In Fig. 9.1, the refractive indices of OEGs up to a degree of polymerization m, $[H–(OCH_2CH_2)_m–OH]$, = 15 at 25° and 45 °C are plotted together with those for tetrahydrofuran (THF) and chloroform. The refractive index (RI) increases with the increase in the degree of polymerization. The RI of chloroform around 20 °C is close to that of diethylene glycol and its peak disappears or appears on the reverse side when SEC is performed with a RI detector around 20 °C using chloroform as the mobile phase. Figure 9.2 is a plot of RI vs. MW for OEG and PEG at 25 °C. The RI increases significantly up to MW 1000, and becomes relatively constant over MW 2000.

The refractive indices of OEG at 25 °C from $m = 2$ to 15 and several PEGs are listed in Table 9.1 [2]. As the RI n of these polymers is a function of MW, the response of an RI detector must be corrected when MW averages are calculated

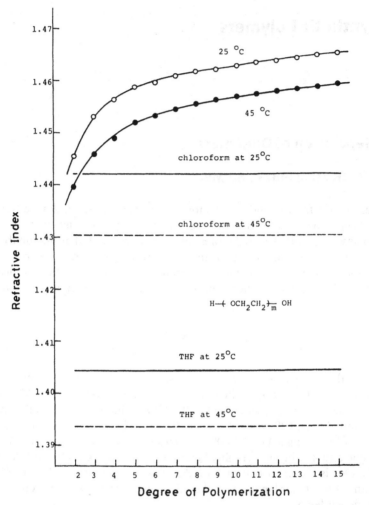

Fig. 9.1. Plot of refractive index vs. degree of polymerization for oligoethylene glycols at 25° and 45 °C

by using an SEC-RI system. (Of course, the response of a UV detector for oligomers is also MW dependent.) The response correction factor for each OEG and PEG when an RI detector is used with THF or chloroform is calculated by the following equation:

$$\text{Response correction factor} = \frac{n_{M=20000} - n_{\text{solvent}}}{n_M - n_{\text{solvent}}} \tag{9.2}$$

where $n_{M=20000}$ is n for PEG of MW 20000, n_M is n for a species of MW M, and n_{solvent} is n for the solvent. The factors at 25 °C are also listed in Table 9.1. The

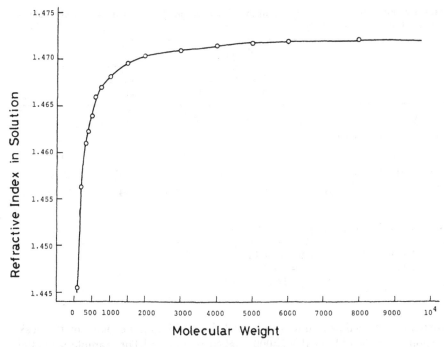

Fig. 9.2. Plot of refractive index vs. molecular weight for oligo- and poly(ethylene glycols) at 25 °C in chloroform

Table 9.1. Refractive indices and response correction factors for oligo- and poly(ethylene glycols) at 25°C [2] [a]

m	n in $CHCl_3$ (25 °C)	Response correction factor		m or MW	n in $CHCl_3$ (25 °C)	Response correction factor	
		in THF	in $CHCl_3$			in THF	in $CHCl_3$
2	1.4455	1.655	8.912	13	1.4645	1.131	1.353
3	1.4529	1.402	2.802	14	1.4650	1.122	1.324
4	1.4563	1.310	2.134	15	1.4655	1.113	1.295
5	1.4589	1.248	1.804	600	1.4660	1.104	1.268
6	1.4597	1.230	1.722	800	1.4674	1.079	1.198
7	1.4610	1.201	1.603	1000	1.4682	1.066	1.161
8	1.4619	1.182	1.530	2000	1.4704	1.030	1.071
9	1.4623	1.174	1.500	4000	1.4715	1.013	1.031
10	1.4630	1.160	1.450	8000	1.4722	1.003	1.007
11	1.4636	1.149	1.409	20000	1.4724	1.000	1.000
12	1.4640	1.141	1.383				

[a] m is the degree of polymerization.

Table 9.2. Effect of response corrections in the calculation of MW averages for several oligo- and poly(ethylene glycols) [2]

	MW average			
	Uncorrected		Corrected	
	M_n	M_w	M_n	M_w
PEG 200 (in THF)	213	230	208	226
PEG 200 (in chloroform)	225	241	208	226
PEG 300 (in THF)	298	323	293	320
PEG 300 (in chloroform)	311	333	297	322
PEG 2000 (in chloroform)	1950	2010	1940	2000
Mixture* (in THF)	414	579	393	550
Mixture** (in chloroform)	468	630	404	572

*, **: Mixtures of PEG 200, 400, 600, and 1000;
     calculated values: * $M_n = 392, M_w = 552$;
                       ** $M_n = 397, M_w = 563$.

correction of each peak intensity is made by multipling the factor by the peak response. The RI of PEG MW 20000 is taken as standard. For example, when the peak intensity of OEG MW 600 is the same as that of PEG MW 20000, then the content of the former is 110% (in THF) or 127% (in chloroform) higher than the latter.

MW averages of several PEG corrected and uncorrected peak responses are listed in Table 9.2 [2]. The difference between the corrected and uncorrected MW averages is smaller in the THF mobile phase than in the chloroform mobile phase, but not noticeably for PEG having a narrow MW distribution. However, mixtures of PEG 200, PEG 400, PEG 600 and PEG 1000 have broad-MW distributions and the effect of the response correction is significant.

9.1.2 Retention Volume and MW

The purposes of oligomer separation by SEC are the determination of MW averages and the quantitative measurement of each component. Recent improvements in SEC columns have enabled complete separations in the MW range of 100–2000. A difference of one carbon atom is sufficient for satisfactory resolution of components in the lower-MW range.

The separation of low-MW compounds and oligomers is based as in the case of polymers on the size of molecules in solution. The molar volume, as well as the chain length and carbon number, can be correlated with the retention volume. Straight lines may be obtained when the retention volume is plotted against log MW in addition to other size parameters. In order to measure MW

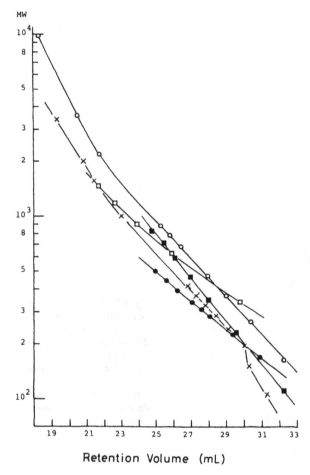

Fig. 9.3. Molecular weight vs. retention volume relationship (1). Column: two Shodex A802; mobile phase: THF; oligomer: (O) oligostyrene, (●) n-hydrocarbon, (□) epoxy resin, (■) p-cresol novolak resin, (×) oligoethylene glycol. Reprinted from Ref. [3] (© 1980) with kind permission from Marcel Dekker Inc., New York, USA

averages, however, a plot of log MW vs. retention volume is more useful than other types of plots.

Similar to polymers (see Chap. 7) the MWs of different types of oligomers having the same size in solution are not always the same. The relationships between retention volume and the log MW for oligostyrenes, epoxy resins, p-cresol novolak resins, OEGs and PEGs, and n-hydrocarbons in THF are shown in Fig. 9.3 [3]. Changes of molecular sizes in different solvents are expected. Figure 9.4 shows a plot of log MW vs. retention volume for the same oligomers in the same column as those in Fig. 9.3, but in a different mobile phase (chloroform). Oligostyrenes, epoxy resins and OEGs, which eluted at the same retention volume as n-hydrocarbons in THF, eluted earlier in chloroform. Epoxy resins and OEGs, which eluted at the same retention volumes as oligosyrenes in THF, eluted earlier in chloroform. p-Cresol novolak resins eluted earlier in THF and later in chloroform than oligostyrenes of the same MW.

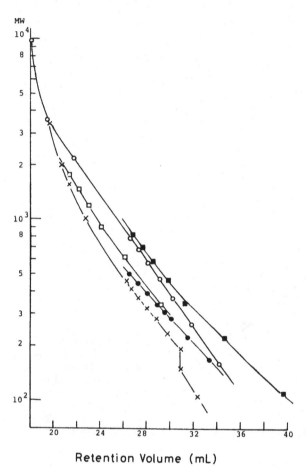

Fig. 9.4. Molecular weight vs. retention volume relationship (2). Mobile phase: chloroform. Other conditions and symbols are the same as in Fig. 9.3. Reprinted from Ref. [3] (© 1980) with kind permission from Marcel Dekker Inc., New York, USA

The gel–solute interactions for oligomers may be large compared to those for polymers and, therefore, oligomers having the same molecular size but different types of chemical structure could elute at different retention volumes. If oligomers are separated solely by size exclusion, the hydrodynamic concept can be applied for the construction of a calibration curve as long as a valid relationship between intrinsic viscosity and MW is established [4]. Intrinsic viscosities for oligomers may be estimated by extrapolations of the Mark–Houwink equation whose parameters are determined from corresponding polymers. However, this approach is not valid for oligomers because the logarithmic plot of intrinsic viscosity vs. MW has a distinct break at MW around 10000 for many polymers and this region should be considered the border between polymers and oligomers. The exponent a approaches the limit 0.5 at very low MWs for oligomers. Therefore, two different universal calibration curves must be constructed at both sides of the border.

Another approach for the construction of the universal calibration curves for oligomers and polymers from MW 1000 to 100000 is to use the Dondos–Benoit equation. This equation is defined as [5]

$$[\eta]^{-1} = -A_2 + A_1 M^{-1/2} \tag{9.3}$$

where A_1 and A_2 are constants related to characteristics of the polymer and of the polymer–solvent pair. Several pairs of A_1 and A_2 for polystyrene (PS), poly-(methyl methacrylate) (PMMA) and poly(vinyl acetate) in several solvents are available in the literature [6]. The parameters of A_1 and A_2 for PS and PMMA in THF with MW ranging from a few hundreds up to 100000 are

PS	A_1	12.3	A_2	0.018
PMMA	A_1	12.1	A_2	0.00877

An equation similar to Eq. (7.48) can be used to calculate the MW of a polymer P under investigation from the MW of polymer S as the standard and is defined as [7]

$$AM_p^{3/2} + BM_p^{1/2} - C = 0 \tag{9.4}$$

$$A = A_{1s} - A_{2s} M_s^{1/2}$$

$$B = A_{2p} M_s^{3/2}$$

$$C = A_{1p} M_s^{3/2}$$

where $A_{1s}, A_{2s}, A_{2p}, A_{1p}$ are the parameters in Eq. (9.3) for polymer S and polymer P, respectively, and M_s and M_p are the MW of polymer S and polymer P.

For calculation of MW averages of oligomers whose chromatograms are not separated perfectly into several peaks, the procedure normally used for polymers can be applied: divide the chromatograms into equal parts as accurately as possible, measure the height at each slice, and calculate MW averages as usual [8]. When a chromatogram is separated into each component, then MW averages can be calculated by knowing the areas S_i and MW M_i of each component by

$$M_n = \sum S_i / \sum (S_i/M_i) \tag{9.5}$$

$$M_w = \sum S_i M_i / \sum S_i \tag{9.6}$$

9.1.3 Epoxy Resins

An epoxy resin is itself a cross-linked polymer and cannot be dissolved in any solvents. Therefore, prepolymers of epoxy resins are candidates for SEC. An epoxy resin consists of molecules which contain more than one 1,2-epoxy group. The epoxide groups of the resin prepolymer react with curing agents to form a cross-linked polymeric structure. The most widely used epoxy resins are based on the diglycidyl ether of bisphenol A, the product of a reaction between bisphenol A and epichlorohydrin. Commercial epoxy resin formulations are complex chemical systems containing a considerable number of components.

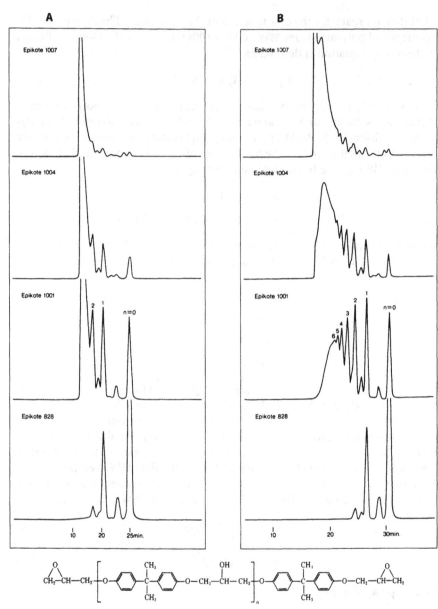

Fig. 9.5 A–B. Chromatograms of epoxy resins [9]. Columns: A Shodex A801 × 2; B Shodex A802 × 2; C Shodex A803 × 2; D Shodex A-804 × 2; mobile phase: THF; flow rate: 1 ml/min; detector: UV (254 nm); column temperature: room temperature

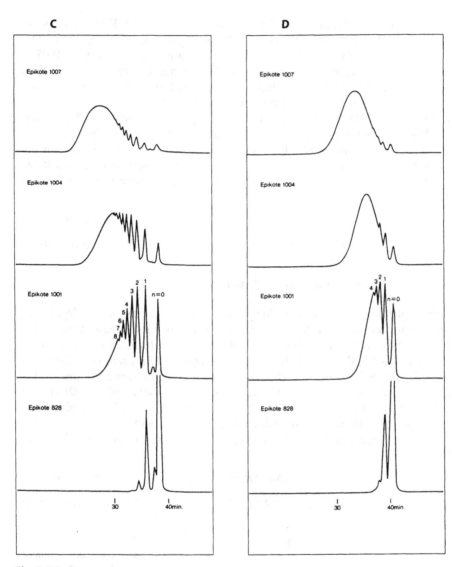

Fig. 9.5C-D

Epoxy resin prepolymers are soluble in most of the solvents that are normally used for SEC. Among them, THF is the most common. Columns for oligomer separation are used for the resins: e.g. Shodex KF801 to 803, TSK gel G1000 to 3000 H_{XL}, Ultrastyragel 100 and 500 Å.

The selection of the pore size of the packing material in the columns is the most important aspect for the high resolution of oligomers. Figure 9.5 shows chromatograms of the commercial expoxy resin prepolymer EPIKOTE obtained with columns of different pore sizes [9]. The Shodex A series is the same as the Shodex KF series in relation to pore size and column designation, but the column length is 50 cm instead of 30 cm for the KF series. EPIKOTE 828 has the lowest MW averages and EPIKOTE 1007 the highest ones among them. The number-average MW of EPIKOTE 1004 is 1470 and that of EPIKOTE 1007 is 3180 by vapor-pressure osmometry. The best resolution for EPIKOTE 828 is obtained with a Shodex A801 column and that for EPIKOTE 1001 with Shodex A802. Shodex A804 has a larger pore size than the other three columns and the resolution for EPIKOTE 828 and 1001 is not as good. The range of MW included in EPIKOTE 1004 is between 340 ($n = 0$) and 8300, but some molecules appear at the exclusion limit of Shodex A802. Shodex A803 is the better column for EPIKOTE 1004. Shoex A803 and A804 columns can be used for EPIKOTE 1007.

The MW relationships between bisphenol-A type epoxy resins and PS in THF are [8, 10]:

$$M_{epoxy} = 4.50\ M_{ps}^{0.748} \qquad \text{(MW range 340–1760)}$$

$$M_{epoxy} = 0.80\ M_{ps}^{0.987} \qquad \text{(MW range 900–}10^5)$$

9.1.4 Phenol-Formaldehyde Resins

Phenol-formaldehyde condensation polymers consist of two types depending on whether acid or base catalyst conditions prevail: novolak type and resol type. Novolak-type resins are soluble in THF and chloroform and separation is performed using the same columns as for oligomers. An example is shown in Fig. 9.6 [11]. Peak 1 is unreacted phenol. Peaks 2, 3, 4 and 5 are as follows in this order: 2,2′-dihydroxydiphenylmethane (DPM); a mixture of 2,4′-DPM and 4,4′-DPM; a mixture of 2, 2′, 6′, 4″-, 2, 2′, 4′, 2″- and 2,2′, 6′, 2″-tetrahydroxytriphenylmethane (trimer); and a mixture of 4, 2′, 4′, 4″- and 4, 2′, 6′, 4″-trimers.

The MW relationships between random-novolak resins and PS in THF are [8, 10]:

$$M_{novolak} = 0.367\ M_{ps}^{1.122} \qquad \text{(MW} \leq 5000)$$

$$M_{novolak} = 0.734\ M_{ps}^{0.955} \qquad (105 \leq \text{MW} \leq 830)$$

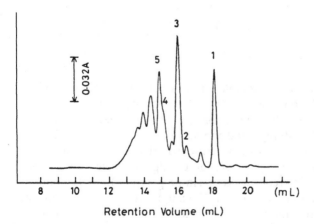

Fig. 9.6. Chromatogram of phenol-formaldehyde novolak resin prepolymers prepared with hydrochloric acid catalyst. Column: two Shodex KF 802; mobile phase: THF; detector: UV at 254 nm; flow rate: 1 ml/min. Reprinted from Ref. [11] (© 1986) with kind permission from Marcel Dekker Inc., New York, USA

Retention Volume (mL)

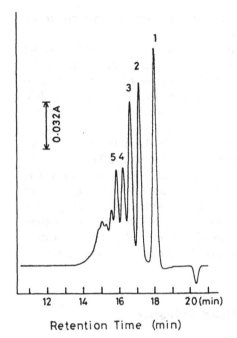

Retention Time (min)

Fig. 9.7. Chromatogram of phenol-form-aldehyde resol resin prepared with alkali catalyst. Separation conditions are the same as those in Fig. 9.6. Reprinted from Ref. [11] (© 1986) with kind permission from Marcel Dekker Inc., New York, USA

A chromatogram of phenol-formaldehyde resols is shown in Fig. 9.7 [11]. Peak 1 is unreacted phenol and peak 2 is 2-methylol phenol (MP). Peak 3 is a mixture of 4-MP and 2,6-dimethylol phenol (DMP). Peaks 4 and 5 are 2,4-DMP and 2,4,6-trimethyl phenol, respectively. Some components in resols are insoluble in chloroform. Besides THF, DMF or 0.1 M aqueous sodium hydroxide can be used as mobile phases with aqueous SEC columns.

9.1.5 Melamine Resins

Melamine-formaldehyde resin prepolymers (methylol melamines) are insoluble in THF, chloroform and other solvents commonly used in SEC. The resins are soluble in DMF or in dimethylsulfoxide. A system of PS gel columns/DMF can be used to separate the resins into individual polynuclear species: mono-, di-, tri-, tetra- and pentanuclear methylol melamines [12].

Alkyl ether derivatives of methylol melamines obtained by the reaction of the methylol group with alcohols are soluble in THF and can be separated with a system of PS gel columns and THF [10]. Silylation of methylol melamines also makes the derivatives soluble in THF [13].

9.1.6 Miscellaneous

There are many examples reported in the literature for the separation of oligostyrenes using SEC columns of narrow pore sizes. Oligostyrenes can serve as a model for the work on the fractionation of oligomers of other polymers. THF is used as solvent for SEC of oligostyrenes.

Minor constituents present in poly(vinyl chloride) were isolated by Soxhlet extraction with diethyl ether followed by SEC with dichloromethane as the mobile phase [14]. Polycarbonate oligomers with up to ten repeating units were separated by SEC with a system of PS gel columns and dichloromethane mobile phase [15]. A mixture of fullerenes C_{60} and C_{70} was separated on PS gel using toluene as the mobile phase [16].

Poly(ethylene terephthalate) (PET) is insoluble in THF, but its oligomers can be dissolved in THF. Therefore, a system of PS gel columns and THF is used to separate PET oligomers [17]. The main component in the chloroform extracts from PET films was a cyclic trimer [18]. The cyclic monomers and oligomers in nylon 6 and nylon 66 were separated and determined with a system of Sephadex gel columns and 0.1 N hydrochloric acid solution [19] and the linear monomers and oligomers were analyzed using a system of Sephadex gel and 0.05 N hydrochloric acid/methanol solution after derivatizing the monomers and oligomers with 2,4-dinitrofluorobenzene [20].

9.2 Polymers with Commonly Used Solvents

9.2.1 Polymer/Solvent Combinations

THF, chloroform, toluene, dichloroethane and methyl ethyl ketone (MEK) are commonly used solvents for SEC at room temperature and they have good compatibility with PS gels. These solvents have adequate properties such as boiling point (50°–110 °C) and viscosity (less than 1 cP). Dichloromethane and p-dioxane can also be used. If the polymer under investigation is soluble in one or some of these solvents, a solvent can be selected as the mobile phase. If not, special solvents have to be used (see Sect. 9.3 or Chap. 10). When an RI detector is used, a solvent whose refractive index is very different from that of the polymer under investigation is recommended in order to get a good response from the detector. Similarly, when a UV detector is used, the solvent should be transparent at the wavelength where the polymer being examined absorbs UV.

Polymer/solvent combinations for use at room temperature are listed in Table 9.3 [21].

9.2.2 Poly(vinyl chloride)

Molecular weight averages of poly(vinyl chloride) (PVC) can be determined using THF as the mobile phase and a refractometer as a detector. Stable aggregates that appear almost universally in PVC solutions prepared at room temperature have been reported to cause problems in the MW measurement of PVC. Heating THF solutions at high temperature for several hours or ultrasonic agitation of the THF solutions may destroy these aggregates. However, these

Table 9.3. Combinations of polymers and commonly used solvents at room temperature [21][a]

Polymer	THF	CHCl₃	toluene	DCE	MC	p-dioxane	MEK
Polystyrene	O	O	O	O	O	O	O
Polybutadiene (polyisoporene)	O	O	O	O	O	O	×
Polyacrylate (polymethacrylate)	O	O	O	O	O	O	O
Poly(vinyl ether)	O	O	O	O	O	O	O
Poly(vinyl alcohol)	×	×	×	×	×	×	×
Poly(vinyl formal (acetal))	O	O	O	O	O	O	–
Poly(vinyl chloride)	O	×	×	×	×	–	O
Poly(vinylidene chloride)	×	×	–	–	–	O	–
Polyacrylonitrile	×	×	×	×	×	×	×
Poly(vinyl acetate)	O	O	O	O	O	O	O
Poly(oxyethylene)	O	O	O	O	O	×	O
Polycarbonate	O	O	O	O	O	–	–
Polyphosphazene	O	×	×	×	×	×	O
Natural rubber	O	O	O	O	O	–	–
Cellulose nitrate (N 12–12.7%)	O	O	×	O	O	–	O

[a] CHCl₃ – chloroform, DCE – dichloroethane, MC – methylene chloride. O: suitable; ×: insoluble; –: no data available.

procedures are hazardous and also problems of molecular degradation may arise during heating or agitation. In addition, the small amount of impurities in PVC such as surfactants used for suspension polymerization decreases the column life time of the PS gel.

Aggregate-free solutions can be prepared in 1,2,4-trichlorobenzene (TCB) by controlling the dissolution time and temperature [22]: PVC solutions are prepared by dissolving known quantities of PVC in TCB at 120 °C for 12 h and then SEC measurements are performed in TCB at 110 °C using PS gel columns. A phenolic antioxidant is added in 0.1% concentration in TCB to prevent oxidative degradation of PVC. Table 9.4 lists MW averages of two PVC samples measured in TCB at 110 °C and in THF at 30 °C for comparison purposes. MW averages measured in TCB are higher than those in THF. Dissolution of aggregates is not complete when PVC is dissolved in TCB at 110 °C, a higher dissolution temperature than 110 °C is required.

Table 9.4. MW averages of PVC measured in TCB and in THF [22]

Sample	Mobile phase, column temp. (°C)	$M_n \times 10^{-4}$	$M_w \times 10^{-4}$	$M_z \times 10^{-4}$
PVC 60 K	TCB, 110	4.39	13.09	74.78
	THF, 30	3.95	10.70	43.97
PVC 66 K	TCB, 110	5.71	16.81	67.36
	THF, 30	4.17	11.06	56.58

9.2.3 Polyorganophosphazene

Careful determination of the MW of polyorganophosphazenes is recommended because of the nature of the substituents, the proportion of residual chlorine atoms after substitution, and any traces of hydrolysis can modify the behavior of these polymers in solution.

MW averages of polyorganophosphazenes based on polycondensation of p-trichloro-N-dichlorophosphonyl monophosphazene can be determined using a system of PS gel columns/THF containing 0.1 M lithium bromide (LiBr) at 30°C [23]. THF is stabilized with 0.03% 2,6-di-tert-butyl-4-methylphenol (BHT). Polyorganophosphazenes are dissolved in THF at 0.2% by gentle stirring at room temperature for a few hours. SEC of poly(diphenoxy) phosphazene ($dn/dc = 0.160$), poly(aryloxy) phosphazene ($dn/dc = 0.145$), and poly(fluoroalkoxy) phosphazene ($dn/dc = -0.029$) can be achieved with this system. For Mark–Houwink parameters of these polymers, see Appendix I.

Poly(diethoxy)phosphazene is determined with a system of PS gel columns/THF including 0.1% tetra(n-butylammonium) bromide [24]. Although poly[bis (trifluoroethoxy)phosphazene] is very soluble in THF, because of the unusual dilution behavior and the high viscosity in THF, cyclohexanone with 0.01 M tetrabutylammonium nitrate is selected as the mobile phase at 40°C [25]. A universal calibration plot is generated using PS, PMMA or poly(tetrahydrofuran) (see Appendix I for Mark–Houwink parameters).

9.2.4 Cellulose Derivatives

Cellulose nitrate (100% nitration) contains 14.1% nitrogen whilst industrial grade materials have lower nitrogen contents within the range of 12.0–12.9%. Although cellulose nitrate containing as little nitrogen as 6.8% is only soluble in water, cellulosics having around 12.7% nitrogen are soluble in THF and chlorinated hydrocarbons. Cellulose nitrate and cellulose tricarbanilate [cellulose tri(N-phenyl carbamate)] are analyzed with a system of PS gel columns/THF at 20°C [26]. Concentrations of the sample solutions are between 0.2 and 1.0% and the injection volumes are between 0.02 and 0.2 ml.

A UV detector can be used for SEC of cellulose nitrates [27]. The Mark–Houwink relationships for the intrinsic viscosity and the degree of polymerization (DP) of cellulose nitrate in THF are as follows [27]:

$$DP < 1000 \qquad K = 0.82 \text{ (ml/g)}; \qquad a = 1.0$$

$$DP > 1000 \qquad K = 4.46; \qquad a = 0.76$$

When cellulose nitrate having a degree of substitution from 2.1 to 2.97% is measured with a system of a silanized silica gel column/THF, polyelectrolyte effects may be observed. The addition of 0.01 M acetic acid to the mobile phase suppresses the non-exclusion effects and leads to validity of a universal calibration between cellulose nitrate and PS [28].

9.3 Polymers Requiring Special Solvents

9.3.1 Polyamides

m-Cresol and o-chlorophenol are good solvents for polyamides and can be used as mobile phases for SEC. As these solvents are viscous, the column temperature must be higher than 100 °C. However, these solvents are also acidic and polyamide samples may be degraded during measurement at high column temperatures. Addition of chloroform or toluene to these solvents is effective in reducing viscosity and for measurements at lower column temperature. Mixtures of m-cresol and chlorobenzene (1:1) or m-cresol and chloroform (35:65) are examples for mobile phases used for polyamides at column temperatures of 40 °C.

Fluorinated alcohols, such as 2,2,2-trifluoroethanol (TFE) and hexafluoro-isopropanol (HFIP), are also good solvents for polyamides and can dissolve them at ambient temperature. The polymer solution shows polyelectrolyte effects that can be avoided by addition of electrolytes. Systems of TFE with 0.05 M LiBr/PS gel columns [29], TFE/silanized silica gel columns [30], HFIP with 0.0005 M sodium trifluoroacetate (NaTFA)/silica gel columns [31] and HFIP with 0.1% NaTFA/PS gel columns [32] have been reported for polyamides (nylon 6, nylon 6, 6, nylon 4, 6, etc.). Insolubility and partial solubility in TFE with 0.05 M LiBr precluded the analysis of nylon 6, 10, nylon 6, 12, nylon 11, and nylon 12 [29].

Fluorinated alcohols are distilled under strict exclusion of humidity, dried over molecular sieves to remove traces of moisture, and degassed by ultrasonic treatment or under vacuum. One of the electrolytes is dissolved in the solvent, followed by filtration through 0.5-μm filters. The polyamide sample is dissolved in the solvent and allowed to equilibrate overnight with slight stirring.

PMMA standards of narrow-MW distribution, which are soluble in the fluorinated alcohols, are used for the calibration of columns instead of the commonly used PS standards which are insoluble in these solvents.

An alternative for room-temperature SEC of polyamides is the use of a mixture of methylene chloride/dichloroacetic acid (80:20, v/v) containing

0.01 M tetrabutylammonium acetate as the mobile phase [33]. PS is soluble in this mixture and therefore PS standards can be used for calibration. Nylon 6, 9, nylon 6, 10, nylon 6, 12, nylon 11, nylon 12 and nylon 6 T, as well as nylon 6 and nylon 6, 6, can be analyzed. Dichloroacetic acid causes skin burns and is corrosive to stainless steel after long-term use. Typically, connecting tube fittings and pump parts must be replaced after 6–9 months of continuous use.

Benzyl alcohol can dissolve polyamides and a system of PS gel columns/benzyl alcohol at 130 °C is applied to conduct SEC of nylon 6, nylon 11, nylon 12 and polyether-b-polyamide [34]. Poly(tetrahydrofuran) standards of narrow-MW distributions are used for calibration.

Trifluoroacetyl derivatives of polyamides, which are prepared by acetylation of polyamide with trifluoroacetic anhydride in methylene chloride at 30 °C, are soluble in THF and other solvents commonly used for SEC [35]. Trifluoroacetylation of polyamides is reproducible, so that routine SEC analysis of N-trifluoroacetylated polyamides is commonly performed. However, the degree of acetylation may result in a difference in the MWs calculated.

9.3.2 Poly(ethylene terephthalate)

Although poly(ethylene terephthalate) (PET) is soluble in m-cresol, because of the high viscosity of m-cresol itself, SEC measurements are performed at elevated temperatures such as 120 °C and the problem of polymer degradation also arises. HFIP is an excellent solvent for PET, and it is soluble at room temperature; however, there are three disadvantages associated with this solvent: health hazards; its high cost and the insolubility of PS in HFIP. The first disadvantage requires the use of good laboratory procedures. Recycling the effluent from the column outlet by redistillation reduces the cost of HFIP [36]. Although PS standards are insoluble in HFIP, PMMA standards of narrow-MW distibution can be used for the construction of a calibration curve. PMMA equivalent MW averages are converted to real PET MW averages by multiplying by 0.57 (the conversion factor) [36].

o-Chlorophenol (OCP) is also used as a solvent for PET. Because of the high viscosity of the solvent, the column temperature must be higher than 90 °C [37]. PS standards can be used for calibration. PET samples are dissolved in OCP at 100 °C within 45 min. A PC calibration curve is converted to a PET calibration curve using the following equation

$$\log M_{PET} = 0.2143 + 0.8975 \log M_{PS}$$

Several mixed solvents consisting of a good solvent for PET and a diluent are reported to perform SEC for PET at room temperature without polymer degradation: HFIP/methylene chloride (3:7, v/v) [38], nitrobenzene/tetrachloroethane (0.5:99.5, v/v) [39], phenol/tetrachloroethane (3:2, w/w) [40], HFIP/chloroform (1:9, v/v) [41] and (2:98, v/v) [42], and dichloroacetic acid/methylene chloride (20:80, v/v) with 0.01 M tetrabutylammonium acetate [43]. PS is dissolved in these mixed solvents and, therefore, a PS calibration curve can be used for the calibration of MW averages of PET. Pentafluorophenol (PFP) can dissolve PET at 60 °C (an elevated temperature is required since it is a solid at rood temperature) and a mixture of HFIP/PFP (1 : 1, v/v) is also used for SEC of PET with no signs of polymer degradation [44].

9.3.3 Other Polymers

N,N-Dimethylformamide (DMF) is a good solvent for poly(acrylonitrile) (PAN), poly(vinyl alcohol) (PVA), PVC, polyarylsulfonate and Trogamide T [poly(trimethyl hexamethyleneterephthalamide)]. Of these polymers, PAN and PVC are also soluble in dimethylsulfoxide (DMSO).

When PAN is separated with DMF alone as the mobile phase, the chromatogram starts from the position of the exclusion limit of the column or earlier as shown in Fig. 9.8a. Portions that elute at or before the exclusion limit may be attributed to the formation of a supermolecular structure such as aggregation or association of molecules. The addition of a small amount of LiBr in DMF eliminates early elution and shows a reasonable chromatogram as shown in Fig. 9.8b [45]. The elution behavior of poly(vinyl pyrrolidone) is similar to that of PAN. PS is soluble in DMF but a retardation of elution is observed when PS gel columns are used and, therefore, the use of poly(ethylene oxide) standards is recommended.

Poly(vinyl butyral) (PVB) is the product of the simultaneous hydrolysis of poly(vinyl acetate) and acetalization of the PVA thus formed with butyraldehyde, producing a copolymer of vinyl alcohol and vinyl butyral. PVB is soluble in THF, but the presence of aggregation in a variety of thermodynamically good solvents such as THF is observed. Aggregate-free SEC is performed with HFIP containing 0.08 % sodium trifluoroacetate as the mobile phase at 45 °C [46].

Poly(vinyl pyridine) is measured with a system of silica-based deactivated columns/pyridine at 25 °C or *N*-methylpyrrolidone at 80 °C [47]. PS is used for

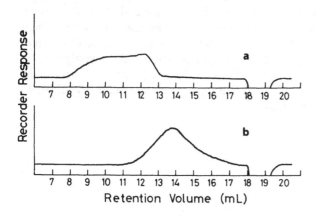

Fig. 9.8. SEC chromatograms of polyacrylonitrile obtained in *a* DMF and in *b* 0.01 M LiBr-DMF at 60 °C. Reprinted from Ref. [45] (© 1983) with kind permission from American Chemical Society, Washington, DC, USA

the calibration curve and universal calibration is effective for this system and samples. For Mark–Houwink parameters, see Appendix II.

Polyamic acid (PAA) and polyamic ester are polyimide precursors and measurement of the MW of the polyimide precursors is complicated due to polyelectrolyte effects such as strong polymer–solvent and electrostatic interactions. In order to suppress the polyelectrolyte effects, a mixture of 0.03 M LiBr, 0.03 M H_3PO_4, and 1 % THF in DMF or 0.03 M LiBr, 0.03 M H_3PO_4, and 1 % THF in dimethylacetamide is used [48]. N-Methylpyrrolidone (NMP) is also used as a mobile phase [49]. In order to remove basic impurities, NMP should be distilled over phosphorous pentoxide (P_2O_5). Mixtures of 0.03–0.06 M LiBr and H_3PO_4 in DMF or 0.06 M H_3PO_4 in DMF/THF (1:1) are recommended mobile phases for both polyamide-imide and PAA [50].

9.4 Selected Topics

9.4.1 Combinations of K and a

There are several combinations of Mark–Houwink parameters, K and a, for a polymer in the same solvent at the same temperature. For example, the number of combinations of K and a for PS in THF at 25 °C listed in Appendix I is 12, ranging from $K = 0.86 \times 10^{-2}$ to 1.76×10^{-2} and $a = 0.679$ to 0.74. The variety of combinations makes it difficult to select an appropriate pair of K and a. An experimentally obtained concept for the combination is that there is a proportional relationship between $-\log K$ and a: small K with large a.

Figure 9.9 shows the relationship between K and a for PS, PMMA and PVC. Data are taken from Appendix I. Open circle and a solid line are for PS at 25 °C and closed circle and a dotted line are for PS at 30 °C. Correlation coefficients of the linear regression lines for PS at 25 and 30 °C, PMMA at 25 °C, and PVC at 25 °C in THF are: 0.952, 0.971, 0.922, and 0.974, respectively.

Any combination of K and a for a polymer under the same experimental conditions listed in the appendices can be used to calculate the viscosity-average MW of the polymer by knowing the intrinsic viscosity, and the MW average calculated is almost similar irrespective of the combination. However, when universal calibration is considered, the selection of the combination of K and a for the calibration standard (e. g. PS) and for the polymer under investigation is important. Combination that appear in the same literature or a combination for the calibration standard specified with that for the polymer being investigated, as in the appendices, must be used.

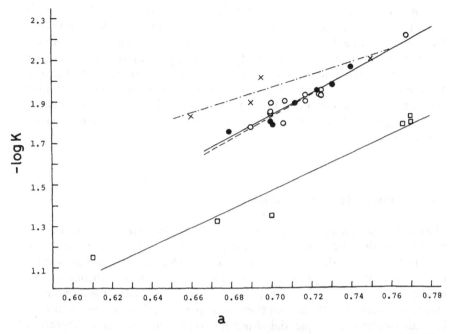

Fig. 9.9. Relationships of Mark–Houwink parameters K and a for PS, PMMA and PVC in THF. $\bigcirc$: PS at 25 °C; ●: PS at 30 °C; ×: PMMA at 25 °C; □: PVC at 25 °C

9.4.2 Calculation of Detector Lag Time

When two or more detectors are used together in series or in parallel to measure MW averages or the chemical heterogeneity of polymers and copolymers, accurate determination of the interdetector volume or lag time between the detectors is very important for matching signals from different detectors. The effective interdetector volume between detectors is not equivalent to the geometric volume because of the difference in the time constant of the detectors, different geometry and volume of the detector cells, and mixing in the cells. The effective interdetector volume is usually obtained experimentally by using a small, monodisperse compound or a polymer standard of narrow-MW distribution. However, in the case of the combination of a MW-sensitive detector, such as a light-scattering detector, and a concentration-sensitive detector, such as an RI detector, the MW at the peak top obtained by the two detectors is different for each detector unless the polymer is truly (molecularly) uniform.

For this reason, several methods for measuring and calculating interdetector volume using a polymeric solute have been implemented in the combination of a low-angle laser light scattering (LALLS) photometer and an RI detector using a PS standard of narrow-MW distribution [51]. Peak onsets of LALLS and RI responses of the standard that were totally excluded from the SEC column gave better precision and accuracy for the measurement of the interdetector volume. The use of a LALLS detector as an absorption spectrophotometer also provided better results. Copper cyclohexanebutyrate was injected into the system without an SEC column and peak onsets of the LALLS detector measured at 632.8 nm and of the RI detector were measured. The best effective value of the interdetector volume was obtained by numerical optimization of the interdetector volume to give the best fit of LALLS local MW of a broad-standard to a narrow-standard calibration curve.

9.4.3 Comparison of Chromatograms

The identity of MW averages determined by SEC can be tested by the t-test by determining the MW averages repeatedly and by knowing the standard deviation. However, even though two polymers have identical MW averages within the experimental error range, a small difference between two chromatograms or MW distributions are often observed. In order to rationalize the difference in chromatograms or MW distributions as experimental variations or the real MW distribution, the sequential U-test for comparing SEC chromatograms of two polymer samples is proposed [52].

Basic parameters for the comparison of two polymers, A and B, are

$$h_0 = -h_1' = -(2\sigma^2/\delta) \ln(1 - 0.5\alpha)/\beta \tag{9.7}$$

and

$$h_1 = -h_0' = (2\sigma^2/\delta) \ln(1 - \beta)/0.5\alpha \tag{9.8}$$

where α is the error of type I, β the error of type II, δ the least difference of the retention volume in the case under investigation, and σ is the standard deviation of the "distinguished point" (dp) values. The dp is defined as the retention volume at 10, 30, 50, 70 and 90% of each integral chromatogram. Parallel measurements of SEC chromatograms of the two polymers are performed in series and the dp values are calculated. The term ΔT_{ij} is defined as

$$\Delta T_{ij} = T_{Aij} - T_{Bij} \tag{9.9}$$

where T_{Aij} and T_{Bij} are the dp values of chromatograms for polymer A and polymer B, the index i defines the number of parallel runs and the index j identifies one of 10, 30, 50, 70, and 90% defined above. After every parallel run, compute ΔT_{ij} for each i and j and summarize the ΔT_{ij} values for i in the case of each j

$$\sum_{i=1}^{n} T_{ij} \tag{9.10}$$

and plot this value at each i.

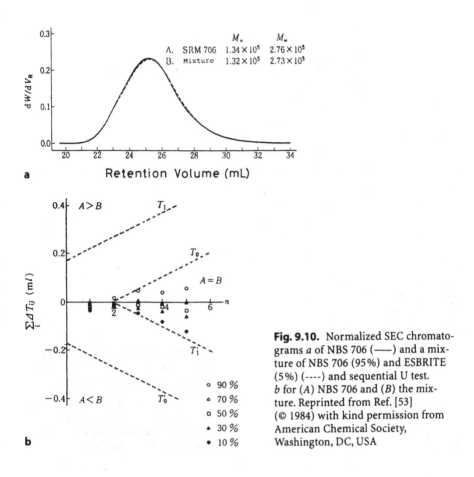

Fig. 9.10. Normalized SEC chromatograms *a* of NBS 706 (——) and a mixture of NBS 706 (95%) and ESBRITE (5%) (----) and sequential U test. *b* for (A) NBS 706 and (B) the mixture. Reprinted from Ref. [53] (© 1984) with kind permission from American Chemical Society, Washington, DC, USA

Examples for the application of the method are shown in Figs. 9.10 and 9.11 [53]. Sample polymers were PS standard NBS 706 (NIST, Washington, DC, USA) and mixtures of NBS 706 and commercial PS ESBRITE (Sumitomo Chemical Co., Tokyo, Japan) (5 and 10%). The parameters were calculated as follows: α is 0.01, β is 0.05, σ is 0.042 ml which was estimated by measuring 20 chromatograms of NBS 706, and δ is 0.1 ml. The value of σ corresponds to 0.3% of the retention volume at the calibration curve of the SEC system and 5% difference of the MW. The broken lines in Figs. 9.10 b and 9.11 b were defined as T_1 (= h_1 + Sn), T_0 (= h_0 + Sn), T_1' (= h_1' − Sn), and T_0' (= h_0' − Sn), where $S = \delta/2$. The two chromatograms shown in Fig. 9.10 a are nearly the same and the MW averages (the mean from three determinations) calculated are almost identical. The results of the sequential U-test shown in Fig. 9.10 b demonstrate that after three or four pairs of runs, all the values of $\sum \Delta T_{ij}$ were found to be located in the area A = B, and that it could be stated at the level of significance of 5% that the two polymer samples had the same MW distribution.

The two chromatograms shown in Fig. 9.11 a are somewhat different, but the MW averages are almost identical. The value of $\sum \Delta T_{ij}$ at j = 10% exceeded the

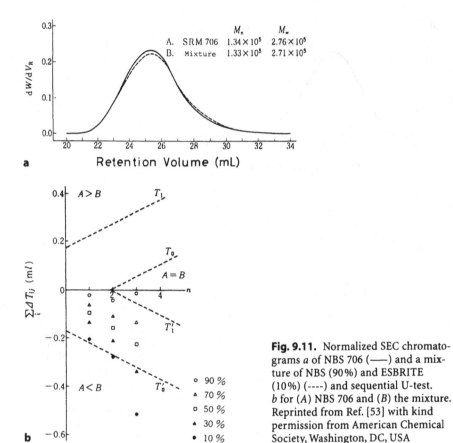

Fig. 9.11. Normalized SEC chromatograms *a* of NBS 706 (——) and a mixture of NBS (90%) and ESBRITE (10%) (----) and sequential U-test. *b* for (*A*) NBS 706 and (*B*) the mixture. Reprinted from Ref. [53] with kind permission from American Chemical Society, Washington, DC, USA

critical value after two pairs of parallel measurements and at the level of significance of 5%, the MW distribution of the mixture was not the same as that of NBS 706 although they had almost identical MW averages.

References

1. WAGNER HL, HOEVE CAJ (1971) J Polym Sci Part A-2 9:1763
2. MORI S (1978) Anal Chem 50:1639
3. MORI S (1980) J Liq Chromatogr 3:329
4. SANAYEI RA, O'DRISCOLL KF (1991) J Macromol Sci Chem A28:987
5. DONDOS A, BENOIT H (1977) Polymer 18:1161
6. TSITSILIANIS C, STAIKOS G (1987) J Appl Polym Sci 33:3081
7. TSITSILIANIS C, MITSIANI G, DONDOS A (1989) J Polym Sci Part B Polym Phys Ed 27:763
8. MORI S (1978) J Chromatogr 156:111
9. Shodex Application Data (1994) (Showa Denko)
10. MORI S (1981) Anal Chem 53:1813

11. MORI (1986) J Liq Chromatogr 9:1329
12. MATSUSAKI T, INOUE Y, OOKUBO T, MORI S (1980) J Liq Chromatogr 3:353
13. BRAUN D, LEGRADIC V (1972) Angew Makromol Chem 25:193
14. DAWKINS JV, FORREST MJ, SHEPHERD MJ (1991) J Chromatogr 550:539
15. BAILLY CH, DAOUST D, LEGRAS R, MERCIER JP, DE VALCK M (1986) Polymer 27:776
16. GÜGEL A, MÜLLEN K (1993) J Chromatogr 628:23
17. MINÁRIK M, ŠIR Z, ČOUPEK J (1977) Angew Makromol Chem 64:147
18. SHIONO S (1979) J Polym Sci Polym Chem Ed 17:4123
19. MORI S, TAKEUCHI T (1970) J Chromatogr 49:230
20. MORI S, TAKEUCHI T (1970) J Chromatogr 50:419
21. FUCHS O, SUHR HH (1975) In: Brandrup J, Immergut EH (eds) Polymer Handbook, 2nd
    edn. Wiley, New York, p IV/241
22. PANG S, RUDIN A (1993) J Appl Polym Sci 49:1189
23. DEJAEGER R, LECACHEUX D, POTIN PH (1990) J Appl Polym Sci 39:1793
24. TARAZONA MP, BRAVO J, RODRIGO MM, SAIZ E (1991) Polym Bull 26:465
25. MOUREY TH, MILLER SM, FERRAR WT, MOLAIRE TR (1989) Macromolecules 22:4286
26. LLOYD LL, WHITE CA, BROOKES AP, KENNEDY JF, WARNER FP (1987) Brit Polym J 19:313
27. MARX-FIGINI M, SOUBELET O (1984) Polym Bull 11:281
28. EREMEEVA TE, BYKOVA TO, GROMOV VS (1990) J Chromatogr 522:67
29. WANG PJ, RIVARD RJ (1987) J Liq Chromatogr 10:3059
30. VEITH CA, COHEN RE (1989) Polymer 30:942
31. SCHORN H, KOSFELD R, HESS M (1983) J Chromatogr 282:579
32. MORI S, NISHIMURA Y (1993) J Liq Chromatogr 16:3359
33. MOUREY TH, BRYAN TG (1994) J Chromatogr A 679:201
34. MAROT G, LESEC J (1988) J Liq Chromatogr 11:3305
35. JACOBI E, SCHUTTENBERG H, SCHULZ RC (1980) Makromol Chem Rapid Commun 1:397
36. MORI S (1989) Anal Chem 61:1321
37. MARTIN L, LAVINE M, BALKE ST (1992) J Liq Chromatogr 15:1817
38. OVERTON JR, BROWING JR HL (1984) In: PROVDER T (ed) ACS Symposium Series No 245,
    ACS, Washington, p 219
39. PASCHKE EE, BIDLINGMEYER BA, BERGMANN JG (1977) J Polym Sci Polym Chem Ed
    15:983
40. UGLEA CV, AIZICOVICI S, MIHAESCU A (1985) Eur Polym J 21:677
41. CHIKAZUMI N, MUKOYAMA Y, SUGITANI H (1989) J Chromatogr 479:85
42. WEISSKOPF K (1988) J Polym Sci Polym Chem Ed 26:1919
43. MOUREY TH, BRYAN TG, GREENER J (1993) J Chromatogr A 657:377
44. BERKOWITZ S (1984) J Appl Polym Sci 29:4353
45. MORI S (1983) Anal Chem 55:2414
46. REMSEN EE (1991) J Appl Polym Sci 42:503
47. RAND WG, MUKHERJI AK (1982) J Chromatogr Sci 20:182
48. WALKER CC (1988) J Polym Sci Part A Polym Chem 26:1649
49. KIM SH, COTTS PM (1991) J Polym Sci Part B Polym Phys 29:109
50. MUKOYAMA Y, SUGITANI H, MORI S (1993) J Appl Polym Sci Appl Polym Symposium
    52:183
51. MOUREY TH, MILLER SM (1990) J Liq Chromatogr 13:693
52. FÜZES L (1979) J Appl Polym Sci 24:405
53. MORI S (1984) In: PROVDER T (ed) ACS Symposium Series No 245, ACS, Washington,
    p 135

10 High-Temperature Size Exclusion Chromatography

10.1 General Procedures for Preparation of Sample Solutions

Several commercially important plastics, such as polyethylene and polypropylene, are not soluble in any solvents at room temperature. They are soluble in o-dichlorobenzene (ODCB) or 1,2,4-trichlorobenzene (TCB) at temperatures above 100 °C and SEC of these polymers is carried out at temperatures above 130 °C. Requirements for solvents used for high-temperature (HT) SEC are high boiling points, i.e. low volatility, good solubility for these polymers and a low viscosity at high temperature, in addition to good compatibility with the stationary phase.

The process of polymer dissolution involves progressive insertion of solvent molecules between the polymer chains, resulting in the swelling of the polymer matrix, followed by separation and disentanglement of the individual polymer molecules. Diffusion of solvent molecules into the polymer matrix is important and, therefore, an increase in temperature increases the rate of dissolution. Increasing the temperature, close to or above the melting point of polymers, assists the dissolution of the polymer in solution. Complete sample dissolution may sometimes be difficult and the insoluble part of the polymer may be present as a microgel. Undestroyed crystallites by incomplete dissolution remain as microgels in solution. Real cross-linked gels are not soluble in any solvents, and chain entanglements in branched polymers are difficult to dissolve and require long periods.

The above considerations are important for the preparation of sample solutions. Sample polymers must be dissolved properly and care must be taken that undissolved microgels are not present in the solution. Procedures for the preparation of polymer solutions depend on polymer types and, in general, samples are dissolved at temperatures 20° to 30 °C higher than the operating temperature. Keeping the sample solution at high temperatures for many hours presents a danger of degradation of the polymer in the solution. However, treatment of the sample solution under mild conditions, such as heating the solvent at the operating temperature for a short time, results in incomplete dissolution of the polymer in solution. The possibility of polymer degradation and the degree of dissolution of the polymer in solution must be checked experimentally by changing the conditions of the preparation of the sample solution. Addition of antioxidants is recommended to reduce oxidative degradation both of sample polymers and the solvent used as the mobile phase. Butylated hydroxytoluene (BHT, 2,6-di-*tert*-butyl-4-methylphenol (Ionol)) and

Irganox 1010 (tetrakis[methylene (3,5-di-*tert*-butyl-4-hydroxyhydrocinnamate)]-methane) are commonly used at concentrations of between 0.05 and 0.1%.

Filtration of sample solutions must be carried out prior to injection as in the case of room-temperature SEC. Solvents used as the mobile phase must be filtered through a membrane of about 0.5 μm. For both convenience and economy, all solvents can be recycled by distilling them under vacuum.

10.2 Selected Applications

10.2.1 Polyethylene

There are three types of polyethylene (PE): high density (HD) PE (linear PE), low density (LD) PE (branched PE), and linear low density (LLD) PE which is a copolymer of ethylene and α-olefins and does not contain long branches. PE must be dissolved at high temperature in ODCB or TCB. It readily comes out of solution on cooling below 100 °C and, therefore, SEC must be carried out at temperatures over 100 °C, preferably 130°–150 °C. TCB is better than ODCB because of its higher boiling and flash point, lower toxicity and higher refractive index. These properties result in higher-temperature operation and greater sensitivity when a refractive index (RI) detector is used.

The crystal melting point of PE is near 140 °C and PE tends to form supermolecular aggregates even at 145 °C in TCB [1]. These structures can be eliminated by heating the solutions to 160°–170 °C for several hours before cooling to the temperature for MW measurements. Higher-molecular-weight HDPE may require longer dissolution times. There are three recommended procedures to avoid poor dissolution and the formation of aggregates.

(1) The sample solutions of PE in TCB are prepared at 170 °C for 1 h without stirring to avoid mechanical degradation, then at 150 °C for 2 h with stirring [2].

(2) The sample solutions are heated at 170 °C for 3 h and the solutions are kept in the heated SEC injector compartment for at least another 3 h prior to injection [3].

(3) Dissolution of PE samples in TCB is achieved by rotating the samples at 160 °C for 16–24 h [4].

SEC measurements are made at 135°–145 °C with a flow rate of 1 ml/min. A temperature of 145 °C is recommended. The injection volume of the sample solution is between 0.1 and 0.2 ml and the concentration of polymer samples is between 0.1 and 0.2%. The absence of aggregates is shown in separate light-scattering measurements by a drastic reduction in large particle "spikes" in light-scattering intensity [1].

An SEC system may be calibrated utilizing a series of polystyrene (PS) standards with narrow-MW distributions and the PS MW at each elution slice can be converted to PE MW using Eq. (7.48). For HDPE in TCB at 145 °C, it will be

$$\log M_{PE} = 0.1760 + 0.9064 \log M_{PS} \qquad (10.1)$$

The values of K and a for HDPE and PS under these conditions are 1.73×10^{-2}, 3.53×10^{-2} (ml/g) and 0.784, 0.617, respectively. For other Mark–Houwink parameters, see Appendix I.

10.2.2 Branched Polyethylene

LDPE is produced by radical polymerization under high temperature and pressure, and the molecules contain both short-chain branches (SCB) and long-chain branches (LCB) at the backbone chain. An SCB distribution as a function of molecular weight is measured using an SEC/RI-infrared (IR) spectrophotometer system. The absorptions of methyl and methylene C–H stretching vibrations which appear at 2965 and 2928 cm^{-1}, respectively, are measured and the number of methyl groups per 1000 carbon atoms is calculated using a relationship between the number and the absorption ratio of both peaks constructed using standard samples [5]. Similarly, the vibrations at 1378 and 1368 cm^{-1} can be used to measure SCBs [6]. LLDPE includes mostly SCBs. Examples are as follows: LLDPE of 1-octene as a comonomer contains 15 SCBs per 1000 carbon atoms at MW = 10^4 and 10 at MW = 10^5 [5]. LLDPE of 1-butene as a comonomer contains an average number of ethyl branches of about 15 per 1000 carbon atoms [6].

LCB distribution cannot be distinguished from SCB distribution by IR spectroscopy. LCB is estimated using a system of SEC/RI-capillary viscometer (CV) or RI-LS. The basic equations and assumptions used in the LCB calculations are:

(1) In SEC, the universal calibration concept that the separation of molecules is controlled by their hydrodynamic volume holds good for both linear and branched PE. That is, when linear (1) and branched (br) PE elute at the same retention volume i, then there is a relationship expressed as

$$[\eta]_{1,i} M_{1,i} = [\eta]_{\mathrm{br},i} M_{\mathrm{br},i} \tag{10.2}$$

In general, $M_{\mathrm{br},i} > M_{1,i}$ and $[\eta]_{\mathrm{br},i} < [\eta]_{1,i}$.

(2) The extent of branching (the branching parameter g) is defined by the ratio of the mean-square radius of gyration of a branched polymer species to that of the linear species of the same MW

$$g = \langle S^2 \rangle_{\mathrm{br}} / \langle S^2 \rangle_1 \tag{10.3}$$

By using the intrinsic viscosity of both polymers, this can be expressed as

$$g^b = [\eta]_{\mathrm{br}} / [\eta]_1 \tag{10.4}$$

where b is a constant, between 0.5 and 1.5, and is related to the structure of branched polymers. The quantities for branched and linear species are taken at the same MW. Alternatively,

$$g^b = (M_{1,i} / M_{\mathrm{br},i})^{a+1} \tag{10.5}$$

where a is the Mark–Houwink exponent and $M_{1,i}$ is not the same as $M_{\mathrm{br},i}$. In this equation, $M_{1,i} / M_{\mathrm{br},i}$ are taken at the same elution volume increment.

(3) The branching index is defined as

$$\lambda = (\text{number of branched points in the molecules, } n_{br})/ \quad (10.6)$$
$$(\text{molecular weight, MW})$$

The branching index is assumed to be independent of MW for a given polymer. Eq. (10.6) is applied to monodisperse polymers. For polydisperse polymers, n_{br} and MW should be n_w and M_w where n_w is the weight-average number of branched points in the molecules and M_w the weight-average MW, respectively.

(4) The number of branched points in the molecule for a randomly branched polymer having trifunctional branched points is calculated from one of the following three equations:

For monodisperse polymers:

$$g_3 = \left[\left(1 + \frac{n_{br}}{7} \right)^{1/2} + \frac{4\, n_{br}}{9\,\pi} \right]^{-1/2} \quad (10.7)$$

For polydisperse polymers:

$$g_3 = \frac{6}{n_w} \left[\frac{1}{2\, n_w^{1/2}} \frac{(2 + n_w)^{1/2}}{} \ln \frac{(2 + n_w)^{1/2} + n_w^{1/2}}{(2 + n_w)^{1/2} - n_w^{1/2}} - 1 \right] \quad (10.8)$$

For fractions of a constant number of branches ($n_{br} > 5$):

$$g_3 = \frac{3}{2} \left(\frac{\pi}{n_{br}} \right)^{1/2} - \frac{5}{2\, n_{br}} \quad (10.9)$$

(5) For tetrafunctional polymers, Eqs. (10.7) and (10.8) become

$$g_4 = \left[\left(1 + \frac{n_{br}}{6} \right)^{1/2} + \frac{4\, n_{br}}{3\,\pi} \right]^{-1/2} \quad (10.10)$$

and

$$g_4 = \frac{1}{n_w} \ln(1 + n_w) \quad (10.11)$$

(6) The weight-average number of long-chain branches per 1000 carbon atoms for the ith slice are given by [7]:

$$\text{LCB}/1000\,\text{C} = (n_{w,i}/M_{br,i}) \times 14\,000 \quad (10.12)$$

An HDPE eluted at the same retention volume as a LDPE has a lower MW than the latter as shown in Fig. 10.1 [8]. Similarly, the intrinsic viscosity of an HDPE is higher than that of a LDPE of the same MW as shown in Fig. 10.2 [9]. The log [η] – log MW relationship for the LDPE is smoothed by a third-degree polynomial regression, though the same relationship for the HDPE is linear.

The branching index of the branched PE, SRM 1476, calculated using Eq. (10.7) was obtained by assuming $b = 1/2$ as follows [10]: $\lambda = 0$ for molecular weights less than 10^4 and λ rises quickly to $(5 - 8) \times 10^{-5}$. The overall branching

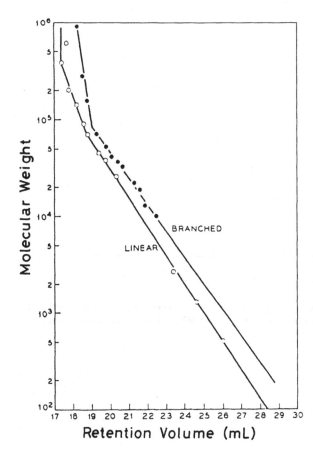

Fig. 10.1. Relationship of molecular weight and retention volume between linear and branched PE. Reprinted from Ref. [8] (© 1967) with kind permission from John Wiley & Sons, Inc., New York, USA

index of the polymer was 2.5×10^{-5} [11]. Another example using Eq. (10.9) for a different LDPE sample showed that λ was about 5×10^{-5} with no significant change along with the MW [9]. A value of $b = 1.2$ was assumed. Relationships of the branching parameter vs. MW and of the branching index vs. MW are plotted in Fig. 10.3. The number of LCBs per 1000 carbon atoms for some LDPE samples increased progressively with decreasing M_w, becoming 10 to 15 at MW $= 10^5$ and almost 0 at MW $= 10^6$ [12]. Ethyl and hexyl side chains do not register as long branches but 16 carbon side chains were counted [13].

The measurement of the MWs of branched PE using an RI-LS system indicated that there was severe shear degradation during the passage of the samples through the SEC columns [14]. It was found that a decrease in the particle size packed in the column from 50 to 10 μm resulted in a decrease in MW of LDPE NBS 1476 from 1.95×10^5 to 1.0×10^5. An increase in the flow rate of the mobile phase from 0.1 to 1.0 ml/min resulted in a decrease in the MW of the same sample from 2.80×10^5 to 1.95×10^5 even in large particle size gel columns. Large particle size columns with low flow rate are recommended for the measurement of MW for branched PE.

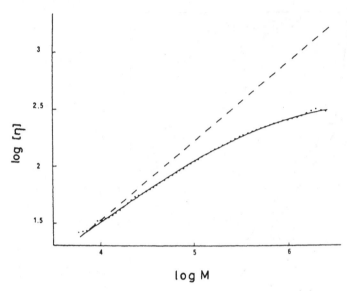

Fig. 10.2. Plot of the intrinsic viscosity for a branched PE (•••) and the comparison with that for a linear PE (---). Reprinted from Ref. [9] (© 1982) with kind permission from John Wiley & Sons, Inc., New York, USA

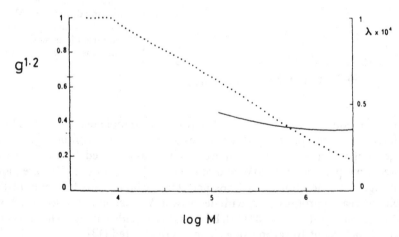

Fig. 10.3. Plot of the branching parameter and the branching index vs. log MW. Reprinted from Ref. [9] (© 1982) with kind permission from John Wiley & Sons, Inc., New York, USA

10.2.3 Polypropylene

Isotactic polypropylene (iPP) is a semicrystalline polymer and stable aggregates cause the same problems for this polymer as they do for PE. Complete dissolution of iPP in solvents is very difficult and, at the same time, iPP is more sensitive to thermal degradation than PE. The highest temperature at which the MW of iPP can be measured by current instruments is 150 °C, but supermolecular aggregates may exist even at this temperature in iPP solutions in TCB and other solvents.

Aggregate-free solutions can be prepared by controlling storage times at 145 °C in stabilized solutions [15]. TCB is used as the mobile phase and is stabilized with 0.1% Irgafos D13-168 (2,4-di-*tert*-butylphenylphosphate) and 0.1% Ionol. Concentrations of iPP solutions are 0.15 – 0.35%. iPP samples are dissolved in the stabilized TCB at 145 °C for several hours and then SEC is performed at the same temperature. The exact time needed for dissolution and measurement varies somewhat with polymer MW. A time of 30 – 50 h from the start is reasonable. Prolonged storage of iPP solutions at 145 °C results in polymer degradation. The effects of storage time of iPP solutions for the dissolution of aggregates and polymer degradation can be followed by measuring the second virial coefficient or the z-average MW with LS. Dissolution procedures at temperatures higher than 145 °C are ineffective with iPP because of thermal instability. The calibration equation for iPP in TCB at 145 °C is, for example,

$$\log M_{pp} = 0.2015 + 0.9187 \log M_{PS}$$

Decahydronaphthalene (decalin) is a good solvent for PP (and PE). When PP is first dissolved in decalin containing antioxidant at 120° – 150 °C, and is then precipitated with acetone or methanol, the solubility of iPP in TCB is increased and can be measured even at 135 °C. The MW values of iPP thus measured are in good agreement with those for samples without precipitation [16].

The MW of iPP can also be measured with a cyclohexane/decalin mixed solvent as mobile phase at 60 °C [17, 18]. Five to ten pellets are dissolved in 1 ml of decalin at 140 °C for about 1 h and then diluted with 9 ml of cyclohexane that have been heated to 75 °C. The solution is then filtered and left in the SEC injector compartment maintained at 60 °C. It is stable for several hours. The column compartment is maintained at 60 °C and a mixture of decalin and cyclohexane (10:90) is used as the mobile phase. Although MW averages obtained by this procedure are somewhat lower than those obtained by the usual procedures (e.g. a iPP sample; $M_w = 1.03 \times 10^5$, $M_n = 2.2 \times 10^4$, $M_w/M_n = 4.68$ by this procedure and $M_w = 1.24 \times 10^5$, $M_n = 3.5 \times 10^4$, $M_w/M_n = 3.54$ by ODCB at 140 °C), this procedure is reproducible and is an attractive alternative to using ODCB and other toxic halogenated aromatic solvents.

10.2.4 Copolymers of Polyethylene

Ethylene-propylene copolymers (EPM) and ethylene-propylene-diene terpolymers (EPDM) are soluble in THF at 30 °C. However, a small amount of lightly cross-linked tree-like structures (microgels) can cause several problems in SEC of these copolymers. Therefore, measurement in TCB or ODCB at temperatures near 140 °C is recommended.

Separation in SEC is achieved according to the sizes of molecules in solution, not to their MWs. Therefore, the retention volume of a copolymer molecule obtained by SEC reflects not only the MW, as in the case of a homopolymer, but simply the molecular size. Copolymers having the same MW but different composition are different in molecular size and elute at different retention volumes. The intrinsic viscosity – MW relationships for copolymers cannot be expressed by a single equation over the entire MW range. The intrinsic viscosity of a copolymer varies not only with MW but with its composition. When the so-called "universal calibration" concept is applied to the MW measurement of copolymers, the composition difference of copolymers must be taken into account.

The Mark–Houwink parameters of EPM copolymers in ODCB at 135 °C are obtained by knowing those of PE, PP and the composition of EPM as follows [19]:

$$a_{EPM} = (a_{PE} a_{PP})^{1/2} \tag{10.13}$$

$$K_{EPM} = K_{PE} W_{PE} + K_{PP}(1 - W_{PE}) - 2(K_{PE} K_{PP})^{1/2} W_{PE}(1 - W_{PE}) \tag{10.14}$$

where W_{PE} is the weight fraction of ethylene in the copolymer and a_{PE}, a_{PP}, K_{PE}, and K_{PP} are 0.74, 0.78, 4.9×10^{-2} (ml/g) and 1.0×10^{-2} (ml/g), respectively.

Ethylene-vinyl acetate copolymer (EVA) with more than 25 % vinyl acetate is soluble in THF, which allows the measurement of EVA copolymers by SEC at room temperature. However, EVA copolymers have LCB in the fractions of high MW and, therefore, it is recommended that EVA is characterized by high-temperature SEC.

An example of measurement is as follows [20]. Dissolution of EVA in TCB is carried out at 135 °C for 4 h in an air oven with gentle, periodic stirring. Four PS gel columns are connected in series and TCB with 1 % poly(ethylene glycol) 400 (PEG 400) is used as the mobile phase at a flow rate of 1 ml/min at 135 °C. The injection volume of the sample solution at a concentration of 0.25 % is 0.4 ml. PEG is added to prevent the adsorption of the vinyl acetate group on the PS gel. Two detectors, RI and CV, are used. Universal calibration can be established by means of a linear PE standard (SRM 1475) and the Mark–Houwink relationship for HDPE, $[\eta]$ (ml/g) $= 0.053 M_v^{0.7}$ (ml/g).

Linear EVA fits the viscosity equation [20]

$$[\eta] = KM^{0.7}$$

$$K = 0.053\,(1 - 0.56\,W_{VA}) \tag{10.15}$$

where W_{VA} is the weight fraction of vinyl acetate and the range of W_{VA} is between 0 and 0.45. For EVA greater than $W_{VA} = 0.45$, the following relationship can be applied:

$$K = 0.053\,(1 - 0.46\,W_{VA} - 0.32 W_{VA}^2) \tag{10.16}$$

This equation offers a good approximation for the K- W_{VA} relationship over the entire range of composition. The relationship between $[\eta]$ and the MW of EVA shows that there is a persistent linear structure for EVA in the low-MW region up to about 4×10^4. Over this MW the value of $g^{1.2}$ decreases linearly from 1.0 to 0.2 at MW = 10^6 and the branching index is approx. 0.4×10^{-4}.

10.2.5 High-Temperature-Resistant Polymers

PPS. Poly(phenylene sulfide) (PPS) is an important engineering plastic that has a high crystalline melting temperature (285 °C). PPS is not soluble in any solvents at temperatures below 150 °C and conventional and high-temperature apparatus cannot be used for the characterization of PPS by SEC. PPS is soluble in 1-chloronaphthalene (CNP) at temperatures over 200 °C; however, serious problems are encountered with the injector valve and detectors. To overcome these difficulties, a new apparatus and procedure are proposed.

A block diagram of the instrument is shown in Fig. 10.4 [21]. The solvent reservoir, the pump and the injector are all operated at room temperature. The preheater consists of 1.5 m × 0.25 mm i.d. stainless-steel tubing and is regulated at 250 °C. The columns are mounted in a column oven which is thermostated at 210 °C. The capillary viscometer consists of capillary tubing and a differential pressure transducer. The capillary is placed in the column oven and the transducer outside the oven.

CNP is purified by passing it through a silica gel column and then filtering it with a 0.5-μm Teflon membrane filter before use. Weighed polymer and solvent (0.2% PSS) are placed in a small pressure container fitted with a valve and a filter holder which is then purged with nitrogen, sealed, and magnetically stirred at 220 °C until dissolution is complete. The container is then quickly inverted and the valve opened. Internal pressure drives the solution into a small

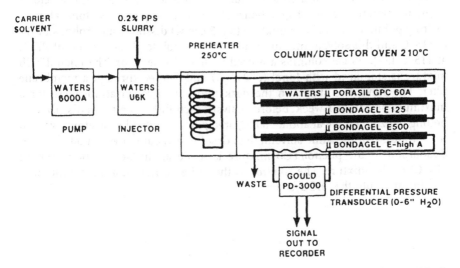

Fig. 10.4. Block diagram of a ultra-high temperature instrument. Reprinted from Ref. [21] (© 1986) with kind permission from John Wiley & Sons, Inc., New York, USA

flask at room temperature to form a finely divided, dilute slurry. A 0.1-ml portion is then injected and the slurry is redissolved by an in-line preheater at 250 °C. Column packings are silica gel and diol-bonded silica gel.

A chromatogram of PPS is obtained as a response of the increase in the pressure drop through the capillary vs. retention volume. Concentration at the increment i relative to the total curve is given by

$$c_i = (d_i/V_i^{a/(a+1)})/\sum (d_i/V_i^{a/(a+1)})$$ (10.17)

where d_i is the recorder deflection above the baseline at i, V_i is the retention volume at i and a is as defined in the Mark–Houwink equation. Universal calibration is carried out using intrinsic viscosity – MW relationships for PS and PPS without using a concentration detector. The disadvantage of this instrument is its low sensitivity to the low-MW fractions.

A modified UV-VIS detector is proposed as an alternaitve to the ultra-high temperature SEC instrument [22]. The cell component is separated from the optical and electrical components and the cell component is heated at 210 °C. To transmit light, an optical fiber system is used from the optical component to the electric one. PS gel columns are used. CNP is opaque below 330 nm and the detector is operated over 356 nm. A flame ionization detector can also be used as the concentration detector [22, 23]. 1-Cyclohexyl-2-pyrrolidinone is transparent above 270 nm and can dissolve PPS at high temperature. This solvent with 20 mM LiCl added is used as the mobile phase and a UV detector is used at 270 nm [24].

PEEK. Poly(arylether ether ketone) (PEEK) is a semicrystalline engineering plastic which is insoluble in all organic solvents at room temperature. PEEK is soluble in some concentrated acids such as sulfuric and hydrofluoric acid at room temperature and also in benzophenone or diphenyl sulfone at near melting point. A phenol/TCB mixture (50 : 50, w/w) dissolves amorphous PEEK at 115 °C and this solvent mixture is used as the mobile phase for SEC of PEEK at 115 °C [25]. PS gel columns are used with an RI detector. The phenol/TCB mixture is stabilized with Ionol (200 mg/l) and filtered through a membrane filter. The solvent is stored in the dark under a nitrogen atmosphere for a maximum period of 1 week. Before analysis, each PEEK sample is dissolved in boiling benzophenone at about 300 °C, precipitated in distilled acetone, washed three times with the same solvent, then dried in a vacuum oven at 60 °C. PEEK dissolution in the phenol/TCB mixture is made at the boiling point (about 187 °C) of the mixture. The solution is then filtered hot on a membrane filter and injected into the SEC system.

References

1. GRINSHPUN V, O'DRISCOLL KF, RUDIN A (1984) J Appl Polym Sci 29:1071
2. LESEC J, MILLEQUANT M, HAVARD T (1994) J Liq Chromatogr 17:1029
3. JIENG L, BALKE ST, MOUREY TH, WHEELER L, ROMEO P (1993) J Appl Polym Sci 49:1359
4. PANG S, RUDIN A (1992) Polymer 33:1949
5. HOUSAKI T, SATOH K (1988) Makromol Chem Rapid Commun 9:525
6. SPRINGER H, HENGSE A, HINRICHSEN G (1990) J Appl Polym Sci 40:2173
7. NEFF BL, OVERTON JR (1982) Am Chem Soc Polym Prepr 23:130
8. SALOVEY R, HELLMAN MY (1967) J Polym Sci Part A-2, 5:333
9. LECACHEUX D, LESEC J, QUIVORON C (1982) J Appl Polym Sci 27:4867
10. WAGNER HL, McCRACKIN FL (1977) J Appl Polym Sci 21:2833
11. WILD L, RANGANATH R, BARLOW A (1977) J Appl Polym Sci 21:3331
12. DAYAL U (1994) J Appl Polym Sci 53:1557
13. RUDIN A, GRINSHPUN V, O'DRISCOLL KF (1984) J Liq Chromatogr 7:1809
14. DEGROOT AW, HAMRE WJ (1993) J Chromatogr 648:33
15. GRINSHPUN V, RUDIN A (1985) J Appl Polym Sci 30:2413
16. LEHTINEN A, VAINIKKA R (1989) International GPC Symposium Newton, MA 612
17. YIN Q, XIE P, YE M (1985) Makromol Chem Rapid Commun 6:105
18. IBHADON AO (1991) J Appl Polym Sci 42:1887
19. OGAWA T, INABA T (1977) J Appl Polym Sci 21:2979
20. LECACHEUX D, LESEC J, QUIVORON C, PRECHNER R, PANARAS R, BENOIT H (1984) J Appl Polym Sci 29:1569
21. STACY CJ (1986) J Appl Polym Sci 32:3959
22. HOUSAKI T (1991) J Appl Polym Sci, Appl Polym Symp 48:75
23. KINUGAWA A (1987) Koubunshi Ronbunshu 44:139
24. MAEDA S, NAGATA M (1993) J Appl Polym Sci, Appl Polym Symp 52:173
25. DEVAUX J, DELIMOY D, DAOUST D, LEGRAS R, MERCIER JP, STRAZIELLE C, NIELD E (1985) Polymer 26:1994

11 Aqueous Size Exclusion Chromatography

11.1 Non-Size Exclusion Effects

11.1.1 Thermodynamics

The determination of molecular weight (MW) averages and molecular weight distributions (MWD) of water-soluble polymers, such as nonionic and ionic synthetic polymers, proteins and peptides, by aqueous SEC sometimes encounters difficulties because of non-size exclusion effects. SEC is a separation technique based on molecular size and any polymers eluted at the same retention volume should have the same molecular size making it possible to calculate the MW or MWD of a polymer using a calibration curve constructed with polymer standards that are different from the polymer under investigation. Moreover, it can be assumed that species that elute earlier from SEC columns have higher MWs than those of the same type of polymer eluted later.

The distribution coefficient in SEC, K_{SEC}, can be defined as

$$K_{SEC} = K_D K_P \tag{11.1}$$

where K_D is the distribution coefficient for ideal size exclusion and K_P is the distribution coefficient for solute–gel (packing materials) interaction effects. In terms of thermodynamic properties, K_{SEC} comprises enthalpic and entropic contributions as

$$K_{SEC} = \exp(-\Delta G^\circ/RT) = \exp(\Delta S^\circ/R)\exp(-\Delta H^\circ/RT) \tag{11.2}$$

where ΔG° is the change in the Gibbs free energy of the system, ΔS° the change in entropy, and ΔH° the change in enthalpy. ΔS° is related to the loss of conformational entropy when 1 mole of solute passes from V_o to V_i and ΔH° is the energy released when 1 mole of solute interacts with the packing (see Chap. 2).

When $\Delta H = 0$, no energetic interactions between the solute and the packing are observed and the separation is solely performed by size exclusion. Therefore,

$$K_D = \exp(\Delta S^\circ/R) \tag{11.3}$$

and

$$K_P = \exp(-\Delta H^\circ/RT) \tag{11.4}$$

11.1.2 Types of Non-Size Exclusion Effects

Intermolecular electrostatic interactions between the solute and the packing (ion exchange, ion exclusion and ion inclusion), intramolecular electrostatic interactions, and adsorption (hydrogen bonding and hydrophobic interactions between the solute and the packing) are the non-size exclusion effects mostly encountered in aqueous SEC [1]. Other non-size exclusion effects such as viscous fingering and concentration effects are not discussed here (see Chap. 5).

(a) **Ion Exchange.** Silica gel has silanol groups which can be dissociated to anionic groups, depending on pH. Silica-based packings such as DIOL-silica, which is glycerylpropyl-bonded silica, and polymer-based packings have residual silanols or carboxyl groups on their surfaces, respectively. These groups are anionic and act as cationic exchange sites. Cationic polyelectrolytes are adsorbed by ion exchange, and anionic polyelectrolytes are excluded from entering the pores of the packings because of electrostatic repulsive forces (ion exclusion).

Ion exchange between anionic packings and cationic polyelectrolytes is an enthalpic interaction (ΔH is negative) and is observed as no elution or elution after the total volume, V_t, of the mobile phase in the column. Methods to eliminate the ion-exchange effect are

- reduce the pH of the mobile phase to below 4 to suppress dissociation of silanol and carboxylic groups,
- add an electrolyte to the mobile phase in an amount of 0.05–0.6 M,
- add a compound (cationic compound for anionic packings) to act in competition with the polymer sample.

(b) **Ion Exclusion.** Electrostatic repulsive forces between anionic packings and anionic polyelectrolytes prevent the solutes from entering the pores of the packings, which is indicated by elution at the exclusion limit, V_o, of the column system or by early elution than expected from MW. Ion exclusion is an enthalpic interaction and ΔH is positive.

Ion exclusion may be eliminated by

- reducing the pH of the mobile phase to below 4,
- adding an electrolyte to the mobile phase in an amount of 0.01–0.2 M.

(c) **Ion Inclusion.** Ion inclusion is caused by the establishment of Donnan membrane equilibrium. During elution, polyelectrolyte counterions have the potential to diffuse freely into the pores of the packings ($K_D = 1$), while the polyelectrolyte is size excluded ($K_D < 1$). Because electroneutrality must be established between the species in the pores and those in the interstitial volume, additional polymer is forced into the packing (ion inclusion) to relax the chemical potential difference. As a result, the elution of polyelectrolytes may be retarded more than expected.

Ion inclusion is indicated by the appearance of an electrolyte peak when a refractive index (RI) detector is used. Methods to eliminate the ion inclusion effect are:

- add an electrolyte to the mobile phase,
- add a low pore size packing to the column set to separate the salt peak from the polymer envelope,
- use a UV detector.

(d) **Intramolecular Electrostatic Interaction.** Because of fixed charges on polyelectrolytes, electrostatic repulsive forces among neighboring ionic sites expand the polymer chains, increasing its hydrodynamic volume. With the addition of an electrolyte to the solvent, these electrostatic forces are screened and the polymer contracts. For example, the intrinsic viscosity of carboxymethyl cellulose is decreased from 8000 to 460 ml/g when the ionic strength is increased from close to zero to 0.7 M [2].

Polyelectrolyte contraction may be followed by determining the K_{SEC} as a function of the mobile phase ionic strength.

(e) **Adsorption.** There are three mechanisms by which a solute can be adsorbed onto the surface of a packing: ion exchange, hydrogen bonding and hydrophobic interaction. Cationic polyelectrolytes are adsorbed onto anionic packings. Details are given in (a) above. Hydrogen bonding can be eliminated by the addition of guanidine or urea to the mobile phase. Hydrophobic interaction is the interaction between the hydrophobic parts of the solute and the packing. For example, the linear backbone of sodium poly(styrene sulfonate) may adsorb onto the hydrophobic parts of glycerylpropyl-bonded silica gel. A reduction in ionic strength or the addition of a surfactant or organic compound, such as alcohol or glycol to the mobile phase, may eliminate this effect.

11.1.3 Elution Behavior of Water-Soluble Polymers

(a) **Non-ionic Polymers.** Non-size exclusion effects between the stationary phase and nonionic polymers, such as poly(ethylene glycols), poly(vinyl alcohol), methyl cellulose and oligosaccharides, are small and the considerations needed for separation procedures are similar to those for non-aqueous SEC.

Polyacrylamide, polyvinylpyrrolidone or starch hydrolysates ocassionally exhibit interactions with packings when pure water is used as the mobile phase. These interactions are due to the basic characteristic of the N-related groups in the polymer chain that act as cationic polymers or are due to the residual carboxylic groups that act as anionic polymers. Therefore, interactions of these solutes are exhibited as adsorption (elution after V_t) or ion exclusion (elution at V_0).

(b) **Anionic Polymers.** Sodium poly(styrene sulfonate) (NaPSS) is a typical ionic water-soluble synthetic polymer. The elution behavior of the polymer is dependent on the ionic strength of the mobile phase and the type of packing used [3]. An example is shown in Fig. 11.1 which shows the universal calibration plots for NaPSS (MW range 4000–177000) and pullulan (a linear polysaccharide) on porous-glass packings (FPG) at various ionic strengths. Pullulan is a nonionic polymer which is independent on the ionic strength for elution. The intrinsic viscosity of NaPSS in the mobile phase increases with a decrease in the ionic strength because of intramolecular chain expansion. NaPSS elutes near the exclusion limit at the lowest ionic strength (3.11×10^{-3} M). Retention volumes of NaPSS increase with increasing ionic strength, approaching the curve for pullulan, but they do not cross the pullulan curve even at ionic strengths as large as 1.35 M. Early elution is due to the ion-exclusion effect between dissociated silanol groups and NaPSS molecules. This effect decreases with increasing ionic strength and "ideal SEC" can be achieved.

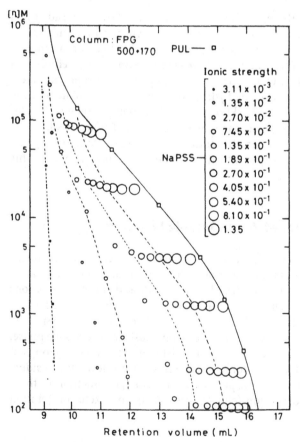

Fig. 11.1. Universal calibration plots for pullulan (□) in pure water and NaPSS (○) in phosphate buffer solutions (pH 8.0) at various ionic strengths. Column: porous-glass (FPG) 500 Å + 170 Å (both in stainless-steel columns of 25 cm × 7.2 mm i.d.); MW range of NaPSS 4000–117000. Reprinted from Ref. [3] (© 1989) with kind permission from American Chemical Society, Washington, DC, USA

Similar plots for a glycercylpropyl-bonded silica packing (e.g. Shodex PRO-TEIN WS-803) are shown in Fig. 11.2. Early elution of NaPSS at the exclusion limit is due to the ion-exclusion effect between the residual dissociated silanol groups of the packing and NaPSS; the ion-exclusion effect decreases with increasing ionic strength of the mobile phase. However, in contrast to the porous-glass packing, the retention volume of NaPSS bisects the pullulan curve at a value of ionic strength which depends on the MW of NaPSS and continues to increase until NaPSS is retained in the column. Increasing the ionic strength leads to a decrease in the effective distance of the ionic atmosphere (the Debye length) on the surface of the packing and of NaPSS.

Consequently, the interactions between the hydrophobic parts of the glyceryl-propyl groups of the packing and of NaPSS increase with an increase in ionic strength. Therefore, the retention volume of NaPSS is governed by a compromise with ion exclusion, size exclusion, and adsorption (hydrophobic interactions). With increasing ionic strength, the ion-exclusion effect is suppressed, but adsorption effects become remarkable.

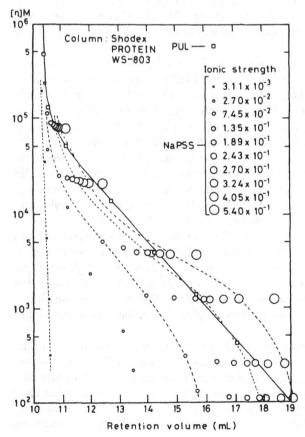

Fig. 11.2. Universal calibration plots for pullulan in pure water and NaPSS in phosphate buffer solutions (pH 8.0) at various ionic strengths. Column: Shodex PROTEIN WS-803 (50 cm × 8.0 mm i.d.). Reprinted from Ref. [3] (© 1989) with kind permission from American Chemical Society, Washington, DC, USA

When columns packed with polymer gel (e.g. Shodex OHpack) were used, a similar pheonomenon to the glycerylpropyl-bonded packing was observed.

(c) **Proteins.** Proteins, which are polyampholytes, consist of α-amino acids that have both carboxyl and amino groups. Neutral amino acids have one carboxyl and one amino group, acidic amino acids have two carboxyl groups and one amino group, and basic amino acids have one carboxyl and two amino groups. When the number of free carboxyl groups and free amino groups in the protein are equal, the isoelectric point is 7.0 and these proteins are termed neutral proteins. The isoelectric point of acidic proteins is lower than 7.0 and these proteins have more free carboxyl groups than free amino groups. Similarly, the isoelectric point of basic proteins is higher than 7.0 and these proteins have more free amino groups than free carboxyl groups.

At the isoelectric point, the number of positive charges (from free amino groups) and negative charges (from free carboxyl groups) in the protein are equal. At pH below the isoelectric point, the number of dissociated amino groups is more than the number of dissociated carboxyl groups and the protein has a positive charge. Similarly, at pH above the isoelectric point, the number of dissociated carboxyl groups is higher than the number of dissociated amino groups and the protein has a negative charge. Therefore, proteins can act as cationic, nonionic, or anionic polyelectrolytes in solution by changing the pH of the solution.

The elution behavior of proteins is dependent on the ionic strength and pH of the mobile phase. Figure 11.3 shows the difference in the retention volume of several proteins with the change in the concentration of phosphate in the mobile phase [4]. Elution behavior can be divided into three groups: (1) retention volume decreases with increasing concentration of phosphate, (2) retention volume is independent of the concentration of phosphate, and (3) retention volume increases with increasing concentration of phosphate. Cytochrome C (isoelectric point, $pI = 10.6$), lysozyme (10.5), trypsin (10.6), and human serum γ-globulin (6.3–8.4), which are basic proteins, belong to the first catagory. Myoglobin ($pI = 7.1$) is a neutral protein and belongs to the second group, and ovalbumin ($pI = 4.6$) and bovin serum albumin ($pI = 4.7$) belong to the third.

Elution behavior is also affected by the pH value of the mobile phase. When the pH value is changed between 4 and 8, the elution behavior of cytochrome C is unchanged because the pI value of the protein is above 8. Myoglobin at pH below 7 behaves like a basic protein such as cytochrome C and at pH 8 behaves like an acidic protein such as ovalbumin. Ovalbumin behaves like a neutral protein such as myoglobin when the pH of the mobile phase approaches the pI.

Besides there being a variation in retention volume with ionic strength of the mobile phase, the shape of the chromatogram also changes. Figure 11.4 shows chromatograms of cytochrome C in different concentrations of phosphate buffer in the mobile phase [4]. Cytochrome C elutes at the total permeation limit ($V_R = V_t = 15.5$ ml) and the shape of the peak is sharp when the concentration of phosphate in the mobile phase is between 0.1 and 0.4 M. The retention

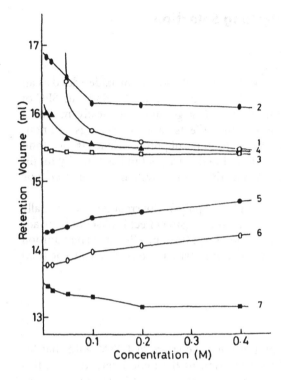

Fig. 11.3. Variation of retention volume of proteins as a function of concentration of phosphate in the mobile phase. Column: glycerylpropyl-bonded porous glass (glyceryl FPG) 500 Å + 170 Å (25 cm × 7.2 mm i.d., × 2); mobile phase: phosphate buffer (Na_2HPO_4 + KH_2PO_4) at pH 7.0 containing 0.2 M NaCl; samples: *1* cytochrome C, *2* lysozyme, *3* myoglobin, *4* trypsin, *5* ovalbumin, *6* bovin serum albumin, *7* human serum γ-globulin. Reprinted from Ref. [4] (© 1987) with kind permision from Marcel Dekker Inc., New York, USA

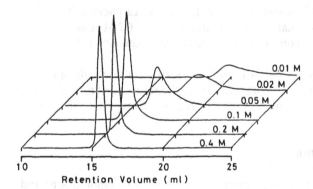

Fig. 11.4. Size exclusion chromatograms of cytochrome C obtained in different concentrations of phosphate buffer used as the mobile phase. Mobile phase: 0.01–0.4 M (Na_2HPO_4 + KH_2PO_4) + 0.2 M NaCl at pH 7.0; column: same as in Fig. 11.3. Reprinted from Ref. [4] (© 1987) with kind permission from Marcel Dekker Inc., New York, USA

volume increases and the peak becomes broad with decreasing concentration of phosphate below 0.1 M. The peak broadening and the increase in retention volume may be due to secondary interactions, such as adsorption between proteins and the packing. Similar effects were observed with changing the concentration of NaCl in the phosphate buffer mobile phase to 0.1 M at pH 7.0.

11.2 Mobile Phase and Packing Selection

11.2.1 Nonionic Polymers

The mobile phase is normally pure water with 0.01 % sodium azide added as an antibacterial agent. Polymer-type packings are preferable for aqueous SEC of nonionic polymers, because unmodified silica gel and chemically modified silica packings sometimes show adsorption effects between packings and nonionic polymers. Distortion of the chromatograms of poly(ethylene glycols) (and oxides) is observed in a system of unmodified silica gel (and chemically modified silica packings)/pure water, while those in a system of polymer-type gels/pure water are symmetrical.

Polar polymers, such as polyacrylamide and polyvinylpyrrolidone, occasionally exhibit non-exclusion effects. When an ion-exclusion effect is observed, the addition of an electrolyte at a concentration of 1 M can be effective. The addition of an organic solvent, such as alcohol or acetonitrile, up to 20 % can also be tried to suppress adsorption.

11.2.2 Anionic Polymers

The convergence of the NaPSS plot at the ionic strength 1.35 M with that for pullulan (Fig. 11.1) confirms the suppression of the ionic interactions at this ionic strength. To a first approximation, a system of unmodified porous-glass or silica gel packing/phosphate buffer at an ionic strength of around 1 M may give "ideal SEC" for NaPSS. When glycerylpropyl-bonded silica packing is used as the stationary phase, phosphate buffer solutions of 0.1–0.2 M ionic strength (Fig. 11.2) can be used as a compromise to help to eliminate ion exclusion, size exclusion, and adsorption.

When hydrophilic polymer gels are used as packing materials, the addition of up to 20 % of an organic solvent, such as methanol or acetonitrile, to the mobile phase can be used to suppress the adsorption effects.

11.2.3 Cationic Polymers

Most LC packing materials have a negative charge, such as silanol groups and residual carboxyl groups, on the surface and act as cation exchangers, so that cationic polymers may adsorb onto the packings irreversibly. One way to minimize this irreversible adsorption is to bond cationic groups chemically to the surface of these packings. These bonded phases provide electrostatic repulsion to polymers of the same charge, thus minimizing adsorption. Similar to anionic polymers with packings having anionic surfaces, the degree of ion-exclusion between cationic polymers and the cationic surface of the stationary phase is controlled by the ionic strength and pH of the mobile phase, although decreased ionexclusion may be offset by increased hydrophobic interactions.

3-Aminopropyl-bonded silica gel can be used as the stationary phase for aqueous SEC of cationic polymers. With this type of packing, several cationic polymers can be separated using mobile phases of 0.1–0.2 M nitric acid, sodium nitrate, a mixture of nitric acid and sodium nitrate, or sodium formate.

Hydrophilic polymer gels used for aqueous SEC also have small amounts of residual anionic groups (mostly carboxyl groups) that act as weak cation exchangers for cationic polymers. Minimization of secondary interactions using these gels is rather easier than with silica-type packings. A mobile phase of 0.1–0.2 M NaCl can minimize the ionic-adsorption effect for cationic polymers. The addition of acetic acid at a concentration of 0.5 M to the aqueous mobile phase containing a small amount of salt (0.1–0.5 M) is also effective to suppress the ionization of the residual carboxyl groups on the surface of hydrophilic polymer gels. The addition of an organic solvent, such as alcohol and acetonitrile, is also a successful strategy to suppress adsorption.

11.2.4 Proteins and Peptides

Mobile phases of phosphate buffer solutions in a concentration of 0.05–2 M at pH 6–7.5 with 0.1–0.2 M NaCl are often used with glycerylpropyl-group-bonded silica gels, commonly used for aqueous SEC of proteins and peptides. A combination of disodium monohydrogen phosphate and potassium dihydrogen phosphate is generally used. The ionic strength and pH of the mobile phase affect the elution of proteins; the former is related to the Debye strength and the latter to ionic dissociation of both sample and the stationary phase. Although so-called "ideal SEC" is a result of a compromise with ion exclusion and adsorption effects, the determination of the solvent conditions needed to optimize size exclusion separation is of practical and fundamental significance. Appropriate conditions of ionic strength and pH are obtained by an empirical optimization procedure to identify the combination of the ionic strength and pH corresponding to the largest value of the regression coefficient, presumably 1.00 [5]. The simplex method is used to maximize the regression coefficient of K_{SEC} in the ionic strength-pH coordinate system. In the system of Superose 6/NaCl–sodium phosphate buffer (9:1) for globular proteins, the optimum conditions were an ionic strength of 0.38 and pH 5.5.

Determination of the MW of proteins using only a calibration curve constructed with standard proteins is not practical because of the possibility of different conformations and also because of secondary interactions with the stationary phase. The use of MW-sensitive detectors together with concentration detectors is practically very important. When proteins that absorb UV radiation are the samples under investigation, then three detectors, UV, RI and LS, can be used to determine the MW of the proteins without having to know the specific refractive index increment dn/dc [6].

$$M = KE^{-1} \, (\text{UV response}) \, (\text{LS response}) \, (\text{RI response})^{-2} \qquad (11.5)$$

where M is the MW of the protein and K is a constant estimated by monitoring a standard protein with a known MW and UV absorption coefficient. E is the UV absorption coefficient expressed in terms of weight concentration of the sample protein.

Glycerylpropyl-bonded silica gel still has adsorption sites. When sample proteins are injected into the column, the proteins that have positively charged basic groups interact with residual silanol groups. The larger the peptides, the less adsorption. Injection of a sufficient quantity of basic proteins or another convenient sample to reduce available adsorption sites before use of the column is recommended [7]. The mobile phase in the column must be displaced with sodium azide (9 ppm)/water or 10% methanol for storage to prevent regeneration of exposed silanol groups.

Trifluoroacetic acid (TFA) is sometimes added to the mobile phase. The advantages of using TFA are increased solubility of proteins, UV absorption at lower wavelengths, and the suppression of silanol dissociation. A concentration

Table 11.1. Characteristics of standard proteins

Protein	MW	pI	Stokes radius (Å)
Cytochrome C	12,500 (13,000)	9.2 (10.6)	16
Lysozyme(egg)	14,300	10.5	
Hemoglobin(human)	16,000	7.0	
Myoglobin	18,000 (17,000)	8.1 (7.1)	20
Trypsin	23,300 (23,000)	8.5 (10.4–10.8)	
Chymotrypsinogen a	25,000 (23,000)	9.2	(22.4)
Pepsin	33,000	2.9	
β-Lactoglobulin	36,000	5.2	
Ovalbumin	45,000	4.6	28
Hemoglobin(bovin)	67,000	7.0	
Bovin serum albumin	67,000	4.7	36
Aldolase	158,000 (148,000)	9.5	(46)
γ-Globulin(human)	160,000	6.3 (6.3–8.4)	
Catalase	240,000 (232,000)	8.0 (5.6)	52
Apoferritin(horse)	467,000	ca 5.0	61
Bovinthyroglobulin	669,000	4.6	85
Ferritin	750,000		

of TFA of 0.1 M (ca. pH 2) suppresses the dissociation of silanol groups (pK_a of a silanol group is 3.5–4).

MWs, pIs (isoelectric points), and Stokes radii of standard proteins are listed in Table 11.1.

11.3 Selected Applications

11.3.1 Synthetic Polymers

(a) **Non-ionic Polymers.** Poly(vinyl alcohols) are measured with a hydrophilic polymer gel/0.05 N $NaNO_3$. Sephacryl S1000 packing material which has large pores up to 400 nm can be utilized to produce highly accurate chromatograms of polymers having molecular hydrodynamic diameters up to 250 nm. Copolymers of sodium-2-acrylamide-2-methylpropane carboxylate and acrylamide, and copolymers of acrylamide and dimethylacrylamide, can also be separated with the system. Partly hydrolyzed polyacrylamides are separated on this packing with 1 M NaCl in water as the mobile phase. Polyvinylpyrrolidone is measured with a system of glycerylpropyl-bonded silica/methanol + water (1:1) + 0.1 M $LiNO_3$. Methanol is more efficient than acetonitrile or dimethyl formamide at suppressing the adsorption effect.

(b) **Anionic Polymers.** Sodium poly(acrylate) can be determined in a system of SEC columns packed with a hydrophilic polymer gel (TSK Gel PW type of Shodex OH type) and 0.3 M NaCl solution as the mobile phase [8]. Poly(ethylene glycols) and poly(ethylene oxides) (PEO) are used as calibration standards.

Fractogel TSK are hydrophilic semirigid spherical gels manufactured from vinyl polymers and designed especially for low-pressure, aqueous SEC. These gels are chemically stable between pH 1 and 13. Universal calibration plots of NaPSS, PEO and dextran in a system of a Fractogel column/0.42 N NaOH mobile phase are in good agreement in a single straight line [9]. Polyacrylamide hydrolysate is measured with a system of Sephacryl S1000/1 M NaCl/UV at 214 nm [10].

(c) **Cationic Polymers.** 3-Aminopropyl-bonded silica gel is a weak cation exchanger and can be used as the stationary phase for aqueus SEC of cationic polymers. However, because of the hydrolytic instability of the cationic stationary phase, separations deteriorate significantly after only a few days. Treatment of the aminopropyl-bonded silica gel with diepoxide improves the hydrolytic stability over many months without any indication of loss of surface coverage [11]. Silica gel was silylated with a 10% solution of aminopropyltriethyloxysilane in refluxing toluene for 15 h under anhydrous conditions. After silylation was complete, epoxide treatment was performed by gently warming the silica in a 5–10% solution of 1,4-butanediol diglycidyl ether in THF for 24 h, followed by treatment with 0.1 M nitric acid for 24 h in order to form the diol. A mixture of poly(2-vinylpyridine) (P2VP) (MW 6×10^5, 9×10^4 and 2.2×10^4) was separated into respective peaks with 0.1 M nitric acid as the mobile phase.

Adenosine diphosphate was separated with the mobile phase of 0.01 M phosphate buffer at pH 7.8. Sodium formate solutions at pH 3.50 from 0.1 to 0.5 M were used as the mobile phases for the separation of P2VP. A mixture of 0.1 M sodium formate at pH 3.50 and 20% methanol was used for the separation of poly[N(1,1-dimethyl-3-imidazolylpropyl)acrylamide] [12]. In the latter case, Mark–Houwink parameters are $a = 0.85$ and $K = 2.98 \times 10^{-3}$ ml/g and those for P2VP in the same mobile phase are $a = 0.95$ and $K = 1.7 \times 10^{-3}$. The mobile phase of 0.2 N $NaNO_3$ + 0.1% trifluoroacetic acid can also be used for the separation of P2VP with aminopropyl-bonded silica gel.

Porous silica gel grafted with quaternary ammonium groups is used as the stationary phase for aqueous SEC of cationic polymers [13]. The universal calibration is generally valid for neutral and cationic polymers in 0.05 M ammonium acetate as the mobile phase when ion exclusion is screened.

Hydrophilic polymer gels can be used for the separation of both anionic and cationic polymers by adding an organic solvent to the aqueous salt solution [14]. P2VP and poly(2-vinylpyridinium bromide) are separated in a mobile phase of 80% (0.5 M Na_2SO_4 + 0.5 M CH_3COOH) + 20% acetonitrile and P2VP is protonated by acetic acid under these conditions. Poly(styrene sulfonate) is separated in a mobile phase of 80% aqueous solution of 0.03 M Na_2SO_4 + 20% acetonitrile. Cationic starch is separated with 0.1 M $NaNO_3$ as the mobile phase.

11.3.2 Biopolymers

(a) **Proteins and Peptides.** Derivatized polystyrene gel in which neutral hydrophilic functionalities are introduced chemically by ether linkages (Rogel-P) is used as the stationary phase and standard proteins are separated with the mobile phase of 70% (95% water + 5% isopropanol + 0.1% TFA) – 30% (95% acetonitrile + 5% isopropanol + 0.1% TFA) [15]. If the latter solution (95% acetonitrile + 5% isopropanol + 0.1% TFA) is less than 15%, some proteins are retained in the column. A small amount of TFA is necessary to elute proteins from the column.

The desalting of peptides, especially when dissolved in phosphate buffers, causes problems, especially for peptides with MW less than 3500, because dialysis is inadequate. The use of hydrophilic-bonded silica gel with the mobile phase of 0.01–0.1 M formic acid is adequate for the desalting of peptides [16]. Formic acid is removed by lyophilization.

Regeneration of the stationary phase is performed with 0.1 M sodium dodecyl sulfate (SDS). The use of volatile aids such as TFA and heptafluorobutyric acid as the mobile phase for the purification of proteins is useful for the preparation of polypeptides free from contaminating salts or detergents [17]. For the separation of a mixture of serum albumin, chymotrypsin, insulin, angiotensin II and met-emkephalin on hydrophilic-bonded silica gel, a solution of 0.02 M TFA is adequate. It is wise to flush the column with water at the end of each working day and store it in 0.02% sodium azide so as not to deteriorate the column.

SDS and guanidine hydrochloride are dissociating and denaturing detergents of proteins. SDS forms a complex with proteins and the structures of the complexes formed between proteins and SDS molecules can be divided into four models: a rod-like particle, a necklace model, a mixed structure of α-helix and a random coil, and a flexible-helix model [18]. An increase in the size of the SDS-protein complex results in a decrease in retention volume, which decreases with an increase in SDS concentration, indicating an increase in size or an elongation of the complex. The final size and shape of the SDS-protein complex

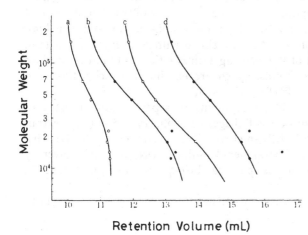

Fig. 11.5. Calibration plots of standard proteins with different concentrations of phosphate buffer and SDS [19]. Column: glycerylpropyl-bonded silica gel; mobile phase: a 0.01 M phosphate buffer +0.1% SDS (pH 7.0), b 0.1 M phosphate buffer +0.1% SDS (pH 7.0), c 0.1 M phosphate buffer +0.02% SDS (pH 7.0), d 0.1 M phosphate buffer + 0.2 M NaCl (pH 7.0)

Table 11.2. Examples of combinations of column packing/mobile phase

Sample	Colum packing/Mobile phase	Ref
Skim milk	Glycerylpropyl-bonded silica/ 0.05 M phosphate buffer (pH 6.8) + 0.1 M Na_2SO_4	20
Peptides	Hydrophilic-coated silica gel/ 20 mM NaH_2PO_4 (pH 6.5) + 0.5% SDS	21
Standard proteins	Hydrophilic-coated silica gel/ 20 mM NaH_2PO_4 (pH 6.5) + 6 M guanidine hydrochloride	21
Standard proteins Whey protein hydrolysates	Superdex-75 HP/125 mM K_3PO_4 + 125 mM Na_2SO_4 (pH 6.65)	22
Immunoglobulin M produced by hybridoma cell culture	Hydrophilic-bonded silica gel/ 0.2 M sodium phosphate + 0.2 M NaCl (pH 6.8)	23

is reached at the critical micelle concentration (CMC) of SDS (ca. 0.18 mM). The change in retention volume of standard proteins with a change in the concentrations of the phosphate buffer and SDS is shown in Fig. 11.5 [19]. Standard proteins used in Fig. 11.5 are cytochrome C, lysozyme, myoglobin, trypsin, ovalbumin, bovin serum albumin and human serum γ-globulin.

Several examples of systems of column packing/mobile phase are listed in Table 11.2 [20–23].

(b) Nucleic Acids. Recombinant DNA technology requires a separation technique that will resolve insert DNAs and DNA linkers from a DNA ligation reaction. A system of hydrophilic polymer gel (TSK G5000PW)/50 mM Tris-HCl buffer at pH 7.4 is used for rapid fractionation of DNA ligation products [24]. Plasmid and fragmented phage DNAs are eluted with recoveries greater than 98%. $MgCl_2$, a component of the DNA ligation reaction, is found to produce DNA-column packing interactions, which are eliminated by chelation with ethylenediaminetetraacetate (EDTA) prior to SEC.

Double-stranded DNA restriction fragments are separated with a system of Suprose 6 (agarose gel of 6% cross-linkage) or Sephacryl S-500 (allyldextran covalently cross-linked with N,N'-methylenebisacrylamide/20 mM Tris-HCl (pH 7.6) + 0.15 M NaCl [25]. The fragment length L of DNA fragments given by the logarithm of the number of base pairs can be expressed as a function of K_{av}

$$L = 2.7623 - 1.9861\ K_{av} \tag{11.6}$$

where K_{av} is defined as

$$K_{av} = (V_R - V_o)/(V_c - V_o) \tag{11.7}$$

The relationship of the MW between rod-like DNA molecules and globular proteins is expressed as

$$M_{DNA} = 101.2\ M_{protein}^{0.495} \tag{11.8}$$

11.3.3 Polysaccharides

Starch hydrolysate (starch syrup) is a mixture of oligosaccharides of different degrees of polymerization and can be separated using columns packed with a cation exchange resins (e.g. Shodex Ionpak KS802) and pure water. A column temperature of 80 °C may be preferable for better resolution. The ionic hydrolysates appear at the exclusion limit with this system. Dextrans are polysaccharides consisting essentially of 1,6-linked D-glucose units with a few percent of branching. Dextrans are obtained as bioproducts of sucrose and are obtained by partial acid hydrolysis to adjust the molecular weight. Therefore, when dextran is measured with a system of hydrophilic polymer gel/pure water, a small peak appears at the exclusion limit, followed by a large main peak. The small peak at the exclusion limit disappears when a hydrophilic polymer gel/0.1 M potassium nitrate is used. The results are shown in Fig. 11.6 [26]. The small

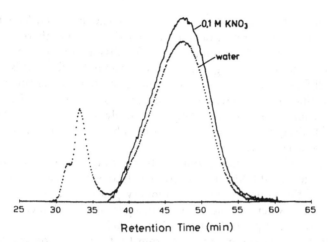

Fig. 11.6. Chromatogram of dextran T500 in water and in 0.1 M KNO₃. Column: Toso PW 6000 + 5000 + 4000; detector: RI. Reprinted from Ref. [26] (© 1987) with kind permission from John Wiley & Sons, Inc., New York, USA

peak also disappears when a 0.01 M NaCl solution is used instead of pure water as the mobile phase. The origin of the small peak is due to the ion-exclusion effect between residual carboxyl groups on dextrans and a negative charge present on the surface of the column packing.

MW averages and MWDs of starch are measured with a system of a column packed with glycerylpropyl-bonded silica gel/dimethylsulfoxide–water (1:3) with 0.01 M potassium phosphate at pH 7.0 as the mobile phase [27]. The addition of potassium phosphate to the mobile phase shifts the chromatograms of oxidized and hydroxyethylated corn starch to higher retention volumes which indicates that the salt reduces the effective hydrodynamic volumes of these starches. The oxidation of starch is known to introduce ionizable carboxyl groups on the starch molecules. The intramolecular expansion by the ionic groups and the ion-exclusion effect between the ionic group and the negative charge on the surface of the packing are lessened by the presence of the salt.

Hemicelluloses (xylose, glucomannan) are a type of polysaccharide present in wood tissues and wood pulp. A system of Separon S HEMA 1000 column packed with poly(2-hydroxyethyl methacrylate-co-ethylene dimethacrylate) gel and 0.5 M NaOH as the mobile phase is used for the determination of MW averages of hemicelluloses [28]. Universal calibration between dextran and xylan is valid.

Native amyloses and hydrolyzed starches are determined with a Superose 6 column (agarose gel)/deaerated distilled water as the mobile phase [29]. Pullulan standards are used as calibration standards. Amylose samples (10–20 mg) are dissolved in 1 ml of 90% dimethylsulfoxide at 60 °C within 4 h. The clear or slightly opalescent solution is mixed with at least four volumes of methanol. The precipitate formed is centrifuged (10 min, 2000 g) and dissolved in 1 ml of water and bubbling steam to remove methanol. The freshly prepared methanol precipitate of amylose is readily soluble in water. The resulting aqueous solutions show retrogradation after about 1 h and, therefore, sample injection must be carried out immediately after dissolution in water.

Heparin is a glycosylaminoglucan having sulfate in the chain. It is a mixture of several different chemical structures and quantitative determination must use several different UV wavelengths. Glycerylpropyl-bonded silica gel/0.1 M NaCl is used for heparin [30]. α-Amylase hydrolysates from corn starch and other alkali hydrolysates of polysaccharides can be measured with a Sepharose column/ 0.2–0.5 N NaOH.

Xanthan, a polysaccharide produced biologically, can be measured with porous silica gel/0.1 M KNO_3 + 1% ethylene glycol which is added to suppress adsorption of the sample onto the stationary phase [31]. Lignin and lignosulfone in waste water are measured with hydrophilic polymer gel/0.025 M ($NaHCO_3$ + NaOH) at pH 10.5 + 0.05% PEG 6000 [32].

Hyaluronic acid (HA) ranges in MW from 10^4 to 10^7 and consists of repeating units of glucuronic acid and N-acetylglucosamine. The MW and concentration of HA present at low levels in biological samples are determined with hydrophilic polymer gel (e.g. Shodex OHpak column)/0.02 M NaCl [33]. A system of Sepharose/0.01 M phosphate buffer (pH 7.4) + 0.15 M NaCl can also be used [34].

References

1. BARTH HG (1987) In: PROVDER T (ed) ACS Symposium Series No. 352, ACS, Washington, DC, chap 2
2. BARTH HG, REGNIER FE (1980) J Chromatogr 192:275
3. MORI S (1989) Anal Chem 61:530
4. MORI S, KATO M (1987) J Liq Chromatogr 10:3113
5. DUBIN PL, PRINSIPI JM (1989) J Chromatogr 479:159
6. MAEZAWA S, TAKAGI T (1983) J Chromatogr 280:124
7. LINK GW Jr, KELLER PL, STOUT RW, BANES AJ (1985) J Chromatogr 331:253
8. KATO T, TOKUYA T, NOZAKI T, TAKAHASHI A (1984) Polymer 25:218
9. CALLEC G, ANDERSON AW, TSAO GT (1984) J Polym Sci Polym Chem Ed 22:287
10. MULLER G, YONNET C (1984) Makromol Chem Rapid Commun 5:197
11. WONNACOTT DM, PATTON EV (1987) J Chromatogr 389:103
12. PATTON EV, WONNACOTT DM (1987) J Chromatogr 389:115
13. DOMARD A, RINAUDO M (1984) Polym Commun 25:55
14. KÜHN A, FÖRSTER S, LÖSCH R, ROMMELFANGER M, ROSENAUER C, SCHMIDT M (1993) Makromol Chem Rapid Commun 14:433
15. YANG YB, VERZELE M (1987) J Chromatogr 391:383
16. RICHTER WO, SCHWANDT P (1984) J Chromatogr 288:212
17. IRVINE GB (1987) J Chromatogr 404:215
18. MASCHER E, LUNDAHL P (1989) J Chromatogr 476:147
19. MORI S, unpublished data
20. DIMENNA GP, SEGALL HJ (1981) J Liq Chromatogr 4:639
21. AHMED F, MODREK B (1992) J Chromatogr 599:25
22. VISSER S, SLANGEN CJ, ROBBEN AJPM (1992) J Chromatogr 599:205.
23. CAURET N, CÔTÉ J, ARCHAMBAULT J, ANDRÉ G (1992) J Chromatogr 594:179
24. HIMMEL ME, PERNA PJ, MCDONELL MW (1982) J Chromatogr 240:155

25. ELLEGREN H, LÅÅS T (1989) J Chromatogr 467:217
26. VAN DIJK JAPP, VARKEVISSER FA, SMIT JAM (1987) J Polym Sci Part B Polym Phys 25:149
27. STONE RG, KRASOWSKI JA (1981) Anal Chem 53:736
28. EREMEEVA TE, BYKOVA TO (1993) J Chromatogr 639:159
29. PRAZNIK W, BECK RHF, EIGNER WD (1987) J Chromatogr 387:467
30. DEVRIES JX (1989) J Chromatogr 465:297
31. LAMBERT F, MILAS M, RINAUDO M (1982) Polym Bull 7:185
32. PELLINEN J, SALKINOJA-SALONEN M (1985) J Chromatogr 322:129
33. MOTOHASHI N, NAKAMICHI Y, MORI I, NISHIKAWA H, UMEMOTO J (1988) J Chromatogr 435:335
34. SHIMADA E, MATSUMURA G (1992) J Chromatogr 627:43

25.

26.

27.

28.

29.

30.

31.

32.

33.

34.

12 Special Applications

12.1 Copolymer Analysis

12.1.1 Introduction

Synthetic copolymers have both molecular weight (MW) and chemical composition distributions. Copolymer molecules of the same molecular size, which are eluted at the same retention volume in SEC, may have different MWs in addition to different compositions. This is because separation in SEC is achieved according to the sizes of molecules in solution, not to their molecular weights. Therefore, determination of accurate values of MW averages and a molecular weight distribution (MWD) for a copolymer by SEC is limited to cases where the copolymer has a homogeneous composition across the whole range of molecular weights.

Determination of the chemical composition distribution (CCD) by SEC has been reported in the literature. Two different types of concentration detectors or two different absorption wavelengths of an ultraviolet (UV) or an infrared (IR) detector are employed and the composition at each retention volume is calculated by measuring peak responses at the identical retention points of the two chromatograms. Details are explained in Sect. 12.1.3. As in the case of MW, molecules that appear at the same retention volume may have different compositions, so that accurate information on chemical heterogeneity cannot be obtained by SEC alone. When the chemical heterogeneity of a copolymer as a function of MW is observed, then the copolymer is said to have heterogeneous composition, but even though it shows constant compositon over the entire range of MW, it cannot be concluded that it has a homogeneous composition [1].

Nevertheless, SEC is still extremely useful in copolymer analysis due to its rapidity, simplicity and wide applicability.

12.1.2 Calculation of MW

A calibration curve for a copolymer consisting of components A and B can be constructed from those of the two homopolymers A and B, if the relationships of the MW and the molecular size of the two homopolymers are the same as their copolymer and if the size of the copolymer molecules in solution is the sum of the sizes of the two homopolymers times the corresponding weight fractions. The MW of the copolymer at retention volume i, $M_{c,i}$, is calculated from the following equation:

$$\log M_{c,i} = W_{A,i} \log M_{A,i} + W_{B,i} \log M_{B,i} \tag{12.1}$$

where $M_{A,i}$ and $M_{B,i}$ are the MWs of homopolymers A and B, and $W_{A,i}$ and $W_{B,i}$ are the weight fractions of components A and B in the copolymer at retention volume i. This equation was empirically postulated for block copolymers [2]. For statistical (random) copolymers, heterointeractions affect the coil size in solution.

The use of the so-called "universal calibration" is a theoretically reliable procedure for calibration. Mark–Houwink parameters for several copolymers are listed in Appendix III.

For ethylene–propylene (EP) copolymers, Mark–Houwink parameters in o-dichlorobenzene at 135 °C are calculated as [3]

$$a_{EP} = (a_{PE} a_{PP})^{1/2} \tag{12.2}$$

$$K_{EP} = W_E K_{PE} + W_P K_{PP} - 2(K_{PE} K_{PP})^{1/2} W_E W_P$$

where W_E and W_P are the the weight fractions of the ethylene and propylene units of the copolymer, respectively.

Calculated Mark–Houwink parameters for poly(styrene–methyl methacrylate) block and statistical copolymers at several compositons in THF at 25 °C are listed in Table 12.1 [4]. The parameters for polystyrene (PS) and poly(methyl methacrylate) (PMMA) used in the calculation are as follows:

PS: $K = 0.682 \times 10^{-2}$ ml/g; $a = 0.766$

PMMA: $K = 1.28 \times 10^{-2}$; $a = 0.69$

If copolymer molecules and PS molecules are eluted at the same retention volume, then

$$[\eta]_C M_C = [\eta]_S M_S \tag{12.3}$$

where M_C and M_S are the MWs of the copolymer and PS, and $[\eta]_C$ and $[\eta]_S$ are

Table 12.1. Calculated Mark–Houwink parameters for poly(styrene–methyl methacrylate) block and statistical copolymers at several compositions in THF at 25 °C [4]

Composition (styrene weight%)	Block copolymer		Statistical copolymer	
	$K \times 10^2$ (ml/g)	a	$K \times 10^2$ (ml/g)	a
20	1.124	0.705	1.044	0.718
30	1.054	0.714	0.953	0.731
40	0.989	0.721	0.879	0.742
50	0.929	0.729	0.821	0.750
60	0.872	0.736	0.779	0.756
70	0.820	0.744	0.747	0.760
80	0.771	0.751	0.722	0.763

the intrinsic viscosities of the copolymer and PS, respectively. A differential pressure viscometer can measure intrinsic viscosities for the fractions of the copolymer and PS continuously, followed by the determination of M_C of the copolymer fraction at retention volume i.

The application of a light-scattering (LS) detector in SEC does not require the construction of a calibration curve using narrow-MWD polymers. However, this method is not generally applicable to copolymers because the intensity of LS is a function not only of MW but also of the specific refractive index (RI) (the RI increment) of the copolymer in the mobile phase. The RI increment is also a function of composition. In the case of a styrene–butyl acrylate (30:70) emulsion copolymer, the apparent MW of the copolymer in THF was only 7% lower than the true one [5]. A recent study concluded that if RI increments of the corresponding homopolymers do not differ widely, SEC measurements combined with LS and concentration detectors yield good approximations to MW and MWD, even if the CCD is very broad [6].

The response obtained from an RI detector may be influenced by the co-polymer composition and must be corrected before the calculation of MW averages. The RI increment $(dn/dc)_{AB}$ for a copolymer AB can be written as

$$(dn/dc)_{AB} = (dn/dc)_A W_A + (dn/dc)_B W_B \qquad (12.4)$$

where $(dn/dc)_A$ and $(dn/dc)_B$ are the RI increments of the corresponding homopolymers A and B, and W_A and W_B are the weight fraction of the A and B units in the copolymer, respectively. The response H_i from the RI detector for the copolymer at retention volume i must be corrected as follows:

$$H_{cor,i} = H_i (dn/dc)_0 / (dn/dc)_i \qquad (12.5)$$

where $H_{cor,i}$ is the corrected response at retention volume i, $(dn/dc)_0$ is the RI increment corresponding to the average composition of the copolymer and $(dn/dc)_i$ is the RI increment of the copolymer at retention volume i.

For poly(styrene(S)–acrylonitrile(AN)) copolymer in THF, the correction for the RI response is as follows [7]:

$$H_{cor,i} = H_i (1 + 0.01 \, AN \, wt\%)$$

and for the P(S-MMA) copolymer in THF the correction is [8]

$$H_{cor,i} = H_i (2.29 - 1.29 \, W_{S,i})$$

where $W_{S,i}$ is styrene wt% in the copolymer at retention volume i.

12.1.3 Chemical Heterogeneity

A point-to-point composition with respect to retention volume is calculated from the two chromatograms obtained using two different concentration detectors or two different wavelengths of an IR detector or a multi-wavelength UV detector. When one of the constituents A or B of a copolymer AB has UV absorption at a specific wavelength and the other is transparent, a UV-RI dual-detector system can be used.

Let A be the constituent that has UV absorption. K_A and K_B are defined as the response factors of an RI detector for the A and B constituents, and K_A' as that of the UV detector for A. These response factors are calculated by injecting known amounts of homopolymers A and B into the SEC dual-detector system, calculating the areas of the corresponding chromatograms, and dividing the areas by the weights of homopolymers injected as

$$F_A = K_A G_A ; \qquad F_B = K_B G_B ; \qquad F_A' = K_A' G_A \qquad (12.6)$$

where F_A, F_B, and F_A' are areas of homopolymers A and B in the RI detector and of homopolymer A in the UV detector, and G_A and G_B are the weights of homopolymers A and B injected into the SEC system. Detector attenuation, flow rate of the mobile phase, and chart speed of the recorder must be adjusted to the basic units or correction has to be made, because response factors depend on these parameters.

The weight fraction $W_{A,i}$ of constituent A at each retention volume i of the chromatogram for the copolymer is given by

$$W_{A,i} = K_B/(R_i K_A' - K_A + K_B) \qquad (12.7)$$

where $R_i = F_{RI,i}/F_{UV,i}$ for the copolymer at retention volume i. Retention volume i for the RI detector is not equal to the retention volume i for the UV detector. Usually the UV detector is connected to the column outlet and is followed by an RI detector and the dead volume between these two detectors must be corrected. The dead volume can normally be measured by injecting a polymer sample having a narrow MWD and by measuring the retention difference of the two peak maxima.

As the additivity of the RI increments of homopolymers is valid for copolymers as in Eq. (12.4), the additivity of the response factors is also valid.

$$K_C = W_A K_A + W_B K_B \qquad (12.8)$$

K_C is the response factor for the copolymer in the RI detector. If the response factors of one or two homopolymers that compose a copolymer cannot be measured because of insolubility of the homopolymer(s), then Eq. (12.8) is employed. Alternatively, the extrapolation of the plot of RI response factors of copolymers of known compositions can be used. An example is shown in Fig. 12.1 where the RI response for PS was 2800 and that for PAN was 2250. The same relationship can be seen in EP copolymers.

Although as mentioned previously the values of these response factors are dependent on several parameters, the ratio of K_A and K_B is almost constant in the same mobile phase. The ratio of PS to PAN in chloroform was 1.244, that of K_{PS} to $K_{polybutadiene}$ was between 1.33 and 1.40.

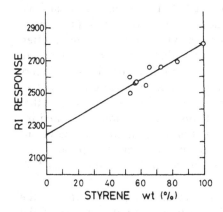

Fig. 12.1. Plot of RI response on composition for poly(styrene–acrylonitrile) copolymers in chloroform

An IR detector can be used at an appropriate wavelength (or wavenumber) to detect one component in copolymers. Information on composition can be obtained by repeating runs and using different wavelengths (wavenumbers) to monitor different functional groups. For styrene–acrylate, styrene–methacrylate, and styrene–vinyl acetate copolymers, a wavenumber around 1730 cm^{-1} for the carbonyl group can be used. Poly(styrene-t-butyl methacrylate) block copolymer in trichloroethylene was measured at three different wavenumbers: 1724 cm^{-1} for the carbonyl group, 697 cm^{-1} for the aromatic group, and 2940 cm^{-1} for the aliphatic C–H group [9]. For P(S-AN) copolymers in chloroform, the wavenumbers at 2222 cm^{-1} for the nitrile group and at 1497 cm^{-1} for the phenyl group were monitored [10] and, similarly, for P(S-MMA) copolymers in chloroform, the absorption at 1730 cm^{-1} for the carbonyl group and at 2950 cm^{-1} for the C–H group.

12.2 Preparative and Recycle SEC

12.2.1 Preparative SEC

The objective of liquid chromatography (LC) is not only quantitative and qualitative determination of components in samples (analytical) but also purification and collection of fractions (preparative). In analytical LC, resolution of chromatograms to obtain accurate and reproducible results is the most important aspect and the sample volume or sample mass injected onto a column should be minimized. However, in preparative LC, recovery of sample mass in a specified time is important and it is often advantageous to apply a much larger sample amount to the column. Although it is generally observed that with increasing sample size (volume or mass) the retention volumes of all compounds and their band widths increase, and resolution becomes worse; a large volume of a sample solution should be injected.

There are two ways to carry out preparative SEC: the use of analytical-scale SEC column(s) and the employment of wider-bore SEC column(s) (preparative SEC column(s)). The scale of conventional analytical SEC columns is often more than adequate. In SEC, it is not expected that synthetic polymers are separated into individual degrees of polymerization. Components of polymers overlap each other and the MW at each retention volume represents only an average value. An increase in sample size results in a decrease in resolution that impedes the determination of narrow MWD polymer fractions. Repeated injections of a smaller amount of a sample solution is required with analytical-scale columns. Oligomers of low MW and proteins, however, are separated into individual peaks and preparative SEC of these materials is similar to small-MW substances.

The number of theoretical plates (N) of a column is proportional to the column length and is independent of the column internal diameter, if the particle size of the packing is the same and the dead volume between a column and a detector is proportional to the column diameter. Therefore, loadability is proportional to the cross-sectional area of the column, provided that the same resolution to cross-section area is maintained. If the column diameter is increased by a factor of 5, then the sample mass or sample volume injected can be increased 25 times without loss of N. Commercially available preparative SEC columns have an internal column diameter of 20 mm and in special cases 50 mm. However, because of the high cost, the particle sizes of the packings are larger than analytical-scale columns and are normally 20 to 30 µm. Therefore, the values of N of preparative columns are somewhat smaller than analytical-scale ones.

Resolution is defined as

$$R_S = 2(V_{R,2} - V_{R,1})/(W_1 + W_2) \qquad (12.9)$$

where $V_{R,1}$ and $V_{R,2}$ are the retention volumes of band 1 and band 2, and W_1 and W_2 are their band widths. The variance of band width is the sum of the column contribution and the equipment contribution that is also the sum of the sample variance, the connecting tube variance, and the cell volume variance (see Chap. 2). The sample variance is proportional to the square of the sample injection volume. This means that the band width increases with an increase in the sample injection volume. The effect of sample mass injected on the band width is somewhat different from that of the sample volume injected. The band width is unchanged up to a certain amount of sample mass injected. Therefore, an increase in sample mass injected is preferable for preparative experiments than an increase in sample volume injected. This consideration is effective for preparative work with oligomers and proteins but not for synthetic polymers because of concentration effects for high-MW polymers.

Provided that two adjacent bands are of equal height and the resolution is 1.25, then the purity of the two fractions cut at the valley of the two bands is 99.4%. By increasing the sample mass or volume injected, the peak intensities of both bands increase and the resolution becomes worse, 0.6 for example. The purity of two fractions cut at the valley becomes 88% (12% contribution from the adjacent band). If the cut point is at the peaks of both bands and the fraction between the two bands is rejected, then the purity becomes 98%, even though the yield of the purified fraction becomes about 60% of that injected. This is one of the ways to get as much material as possible fractionated at a time.

The flow rate of the mobile phase for preparative columns must be increased proportionally to the cross-sectional area of the columns in order to maintain the same linear flow velocity and separation time as in analytical-scale columns. For a 20-mm i. d. column, the flow rate may be increasd to 6 to 7 ml/min and it is preferable to use a pump which can deliver solvent at rates up to 100 ml/min. Similarly, the sample injection volume should be about 1 ml or much larger. In preparative SEC, solute concentrations are generally high and sometimes saturation of peaks is observed: a UV detector with a short-pathlength cell is useful. A shift from wavelength of solute absorption maximum to decrease detection sensitivity reduces the potential for nonlinear detector response.

Contamination during fractionation must be avoided. Tetrahydrofuran (THF) may contain an antioxidant, such as 2,6-di-*tert*-butyl-*p*-cresol and, therefore, it cannot be used as a mobile phase for preparative purposes. THF without an antioxidant can be used if the THF is protected by an inert gas from contact with the atmosphere. In aqueous SEC, mobile phases are often buffer solutions and the inclusion of inorganic electrolytes in fractionated materials is inevitable. In this case, volatile buffer reagents, such as ammonium chloride or a pyridine/formic acid system are employed, so that these electrolytes are removed by evaporation of the fractionated solutions. Impurities of non-chromophores in a UV detector and those having a similar RI to the mobile phase in an RI detector sometimes result in contamination.

12.2.2 Recycle SEC

Peak capacity, defined as the number of components separated in a column at one time, is expressed as

$$m = 1 + 0.2 \, N^{0.5} \tag{12.10}$$

where the ratio of retention volume of the slowest-moving peak to that of the fastest-moving peak is assumed to be 2.3. In reversed-phase LC, this ratio is assumed to be 10 and the value m becomes 2.9 times larger than that in SEC, provided that both columns have the same value of N. In addition, the ratio in reversed-phase LC can be increased by changing the LC conditions, resulting in an increase in the value m. However, the ratio in SEC is a characteristic of the SEC column and cannot be changed by experimental conditions.

One possible way to increase resolution in order to separate closely eluting components completely is to increase the number of columns. Resolution increases linearly with the square root of column length, but the back pressure, the separation time, and cost also increase linearly with length. Therefore, there is a finite restriction on maximum column length. Recycling the sample through the same column set one or more times is equivalent to increasing column length without the need for additional columns: one recycle corresponds theoretically to a two-fold increase in column length and two recyles correspond to a three-fold increase in column length.

For $(n + 1)$ passes through the column (the number of recycles is n), the column resolution becomes

$$R_{s,n+1} = (n + 1)^{0.5} R_s \qquad (12.11)$$

Recycling is usually carried out through the pump and band broadening caused by extra-column effects decreases recycling efficiency. Therefore, Eq. (12.11) is rewriten as

$$R_{s,n+1} = \frac{2(n + 1)(V_{R,2} - V_{R,1})}{[(n + 1) W_1^2 + nW_E^2]^{0.5} + [(n + 1)W_2^2 + nW_E^2]^{0.5}} \qquad (12.12)$$

where W_E is the band width caused by the extra-column effects.

The number of recycles is not infinite. The fastest-moving peak overtakes the slowest-moving peak and remixing occurs. A "draw-off" procedure must be used to eliminate unwanted components to avoid remixing by additional cycles. An example is shown in Fig. 12.2 for polystyrene 600 (MW is 600) [11]. Two JAI GEL 2H columns (600 × 20 mm i.d.) packed with polystyrene gel which had an exclusion limit of MW 8000 were used. The eluent was chloroform at a flow rate of 2.88 ml/min. The sample concentration was 3% and the injection volume was 3 ml. In order to prevent overlapping of the end of the chromatogram in cycle n with the front in cycle $n + 1$, the last parts of the chromatogram in cycle n were collected or removed from the system by turning the recycle valve to the collect position.

In cycle 2 (the first recycle) in Fig. 12.2, the eluate from cycle 1 was returned to the column by switching the recycle valve to the recycle position. Peaks of dimer (peak 2) and trimer (peak 3)(cross-hatched area at cycle 2) were collected by switching the recycle valve to the collect position when the peak of the trimer appeared in the detector (the arrow at cycle 2). After collecting peak 2, the recycle valve was switched to the recycle position. In cycle 3, the recycle valve was switched to the collect position when the peak of the tetramer appeared in the detector (the arrow at cycle 3), and the tetramer and the rest of the trimer were collected, followed by switching the valve to the recycle position again. This procedure was repeated and remixing was prevented.

A schematic diagram of a recycle system is shown in Fig. 12.3 for which the solvent-flow options are indicated by the arrows. The difference from a conventional SEC system is that the recycle system includes a recycle valve between the outlet of the detector (an RI detector in Fig. 12.3) and the inlet of the pump that is connected with a tee (a 3-way joint) to the solvent reservoir and the recycle valve. Figure 12.3 shows the recycle operation (see the position of the recycle valve designated with a solid line) where the sample is passed through the column and the RI detector and back through the pump. No new solvent passes from the solvent reservoir into the column through the pump during the recy-

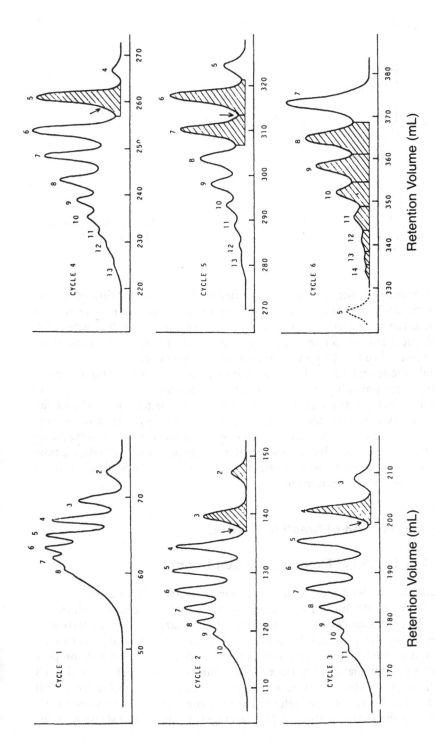

Fig. 12.2. Recycle size exclusion chromatograms of PS 600. Reprinted from Ref. [11] (© 1978) with kind permission from Elsevier Science-NL, Sara Burgerhartstraat 25, 1055 KV Amsterdam, The Netherlands

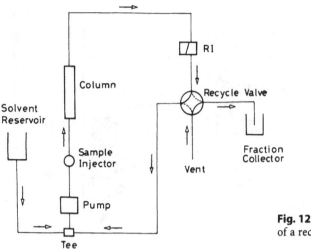

Fig. 12.3. Schematic diagram of a recycle system

cle operation. The collecting tubing is washed by flowing solvent through a vent. When the desired fraction is monitored by the detector, the recycle valve is turned to the collect position (see the path of the recycle valve designated with dotted lines) and collection of the fraction is performed. At the same time, solvent flows through the pump from the solvent reservoir.

Band broadening due to the pump chamber must be carefully minimized. A traditional reciprocating pump is preferable to a modern HPLC pump, because the former has a smaller chamber volume than the latter. The outlet of the solvent reservoir may be connected to the vent, not to the tee, so that it is possible to wash out the residue in the tube between the recycle valve and the pump inlet when the recycle valve is turned to the collect position. Similarly, the tube between the recycle valve and the fraction collection can be washed when the recycle valve is in the recycle position.

12.3 Separation of Small Molecules

12.3.1 Selection of Experimental Conditions

The small molecules discussed in this section are restricted to oligomers having a MW less than 1000 including monomers and low-MW materials which are not polymerized substances. SEC is a technique that separates materials according to their molecular sizes and this concept is still effective for small molecules.

Columns used for the separation of small molecules have narrow pore diameter packing materials. Shodex KF-801, TSK G1000H, PLgel 50 Å and 100 Å and Ultrastyragel 100 Å are used for this purpose. The primary objective of this separation is the quantitative determination of components in the samples, not necessarily the determination of MW averages, so that the resolution of the columns is the most important aspect. Columns with high resolution, in other words, those having a higher number of N must be used. Separability of small molecules is primarily dependent on the pore diameter of the packings in the column as in polymer separation and, therefore, insufficient separation must be supplemented by recycling or connecting many columns.

Selection of the mobile phase is especially important for the separation of small molecules. Interactions between the mobile phase and the stationary phase and between the mobile phase and the sample molecules are more remarkable than those with sample polymers (see Sect. 9.1 and Figs. 9.3 and 9.4). Retardation of small molecules when a polar solvent is used as the mobile phase may be observed. For example alkyl carbonic acids, such as acetic acid, elute at a reasonable retention volume when THF is used as the mobile phase but retard when N,N-dimethylformamide is used. The addition of a polar or a nonpolar solvent to the mobile phase influences predominantly the elution of small molecules. An example is shown in Fig. 12.4 [12]. Retardation of phthalate esters is observed when increasing n-hexane in the chloroform mobile phase. This is because the partition mode superimposes on the size exclusion mode.

When an RI detector is used, attention must be paid to the RI of the mobile phase. As shown in Fig. 9.1, for example, diethylene glycol and chloroform have similar refractive indices at 25 °C and the difference in the relative response factor of diethylene glycol in chloroform to other oligomers is larger than in a

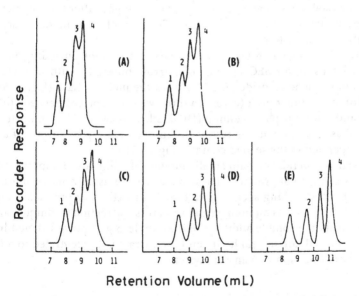

Fig. 12.4. Size exclusion and normal phase partition chromatograms of phthalate esters on a 4.5-mm i.d. column packed with Shodex A801 PS gel. Flow rate: 0.48 ml/min. Mobile phase: *A* Chloroform; *B* chloroform/*n*-hexane (9:1, v/v); *C* (8:2); *D* (7:3); *E* (5:5). Solutes: *1* dinonyl-, *2* dibutyl-, *3* diethyl-, *4* dimethylphthalate. Reprinted from Ref. [12] (© 1979) with kind permission from American Chemical Society, Washington, DC, USA

THF mobile phase. Quantitative experiments using an RI detector sometimes lead to serious errors when the difference between the RI of the solute and that of the mobile phase is small because of the temperature dependency of the RI [13].

12.3.2 Additives in Polymers

Several types of additives are added to polymers to improve their properties and to stabilize them: plasticizers, antioxidants, UV stabilizers, and so on. These additives are separated and analyzed by SEC and other LC separation modes. The size exclusion mode can be used satisfactorily when the molecular sizes of the additives are different, but other separation modes should be used when molecular sizes are similar. For example, a mixture of phthalate esters (methyl-, ethyl-, n-butyl-, n-octyl- and n-nonylphthalate esters) can be separated completely by SEC using the columns described in Sect. 12.3.1, as well as by other separation modes [14].

There are two methods to analyze additives in polymers. One is to extract the additives in the polymers using an appropriate solvent and then separate the extract by SEC or with other separation modes. Plasticizers in poly(vinyl chloride) film (or sheet) are extracted by immersing the polymer in n-hexane overnight or by extracting it with n-hexane using a Soxhlet extractor. The extract is separated by SEC using THF or chloroform as the mobile phase. The other method is to inject the polymer solution into an SEC system which has a column for small molecules, and the peak(s) that appear after the polymer peak is allowed to flow into a second column (SEC or another LC mode) to separate and analyze the additives.

An example is shown in Fig. 12.5 [15]. First, a PS sample including additives is separated using an SEC column of narrow pore size with a mixture of n-hexane/methylene chloride (73:27, v/v) as the mobile phase (Fig. 12.5a). At the end of the polymer peak (arrow), a three-way valve placed between the SEC column and a normal-phase column (DIOL-silica) is switched from waste to the normal phase column and the rest of the compounds, e.g. the group of additives, are separated in the second column (Fig. 12.5b).

Polyolefins generally contain small amounts of additives to improve various properties of the final products. These materials are usually one or more antioxidants, light absorbing agents or colorants, as well as by-products from the polymerization, such as monomers and initiators. Extraction techniques are applied to isolate the soluble additives. For example, 5 g of polyethylene film are immersed in THF for several hours and the extract is concentrated to a factor of 20 prior to injection into an SEC system.

12.3.3 Application as a Cleanup Tool for Residue Analysis

SEC is used as a versatile cleanup method for the isolation of organic contaminants from environmental matrices: the separation of the analytes from co-

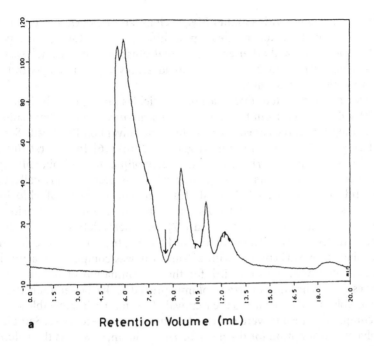

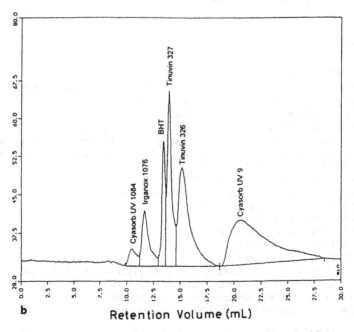

Fig. 12.5. Chromatograms obtained from **a** polymer (first peak) and additives only with an SEC column and **b** final chromatogram of the analysis of additives. Reprinted from Ref. [15] (© 1995) with kind permission from Elsevier Science-NL, Sara Burgerhartstraat 25, 1055 KV Amsterdam, The Netherlands

extracted compounds with a higher MW, such as fats and dyes. SEC is currently used on a routine basis for the cleanup of different agricultural products, animal fats, foods, soils, and other environmental substances in the analysis of organochlorine and organophosphorus pesticides, polychlorobiphenyls and polycyclic aromatic hydrocarbons.

Pesticides (chlorinated hydrocarbon pesticides and polychlorinated biphenyls) are extracted from the sample with a mixture of methylene chloride/cyclohexane (9:1) and cleaned up with the same solvent on Bio-Beads S-X3 gel (PS gel, 50 g in a 400 mm × 25 mm i.d. glass column) [16]. In the first 100 ml of eluate, the interfering matrices, such as phospholipid, fat and oil, chlorophyl, and carotenoid components, are separated, while the pesticides are collected in the second cleaned-up (100–160 ml) fraction. The recovery of chlorinated pesticides is between 90 and 100%. As a substitute for methylene chloride, acetone and petroleum ether can be used [17]: acetone/cyclohexane (25:75) and acetone/petroleum ether (1:1) with 85 g of PS gel in the same glass column.

A miniaturized SEC column with a 2-mm i.d. was compared to a routinely used SEC column with 10-mm i.d. for the determination of organochlorine pesticides [18]. The miniaturized column proved to be better than the conventional column with regard to the removal of the matrix in vegetable samples, while comparable results were obtained with regard to recoveries. Sample size and solvent consumption for the miniaturized column, as well as the volume of the collected fraction, were dramatically decreased.

References

1. MORI S (1987) J Chromatogr 411:355
2. RUNYON JR, BARNES DE, RUDD JF, TUNG LH (1969) J Appl Polym Sci 13:2359
3. OGAWA T, INABA T (1977) J Appl Polym Sci 21:2979
4. GOLDWASSER JM, RUDIN A (1983) J Liq Chromatogr 6:2433
5. MALIHI FB, KUO CY, PROVDER T (1984) J Appl Polym Sci 29:925
6. KRATOCHVIL P, NETOPILÍK M, BOHDANECKÝ M (1995) 8th International Symposium on Polymer Analysis and Characterization (ISPAC-8), May, 1995, Florida, USA, Abstract L14
7. MORI S (1980) J Chromatogr 194:163
8. MORI S, SUZUKI T (1981) J Liq Chromatogr 4:1685
9. DAWKINS J, HEMMING M (1975) J Appl Polym Sci 19:3107
10. MORI S (1982) J Chromatogr 246:215
11. MORI S (1978) J Chromatogr 156:111
12. MORI S, YAMAKAWA A (1979) Anal Chem 51:382
13. MORI S (1983) J Liq Chromatogr 6:813
14. MORI S (1976) J Chromatogr 129:53
15. NERÍN C, SALAFRANCA J, CACHO J, RUBIO C (1995) J Chromatogr A 690:230
16. STEINWANDTER H (1982) Fresenius Z Anal Chem 313:536
17. STEINWANDTER H (1988) Fresenius Z Anal Chem 331:499
18. VAN RHIJN JA, TUINSTRA LGMTH (1991) J Chromatogr 552:517

Appendix I

Mark–Houwink Parameters for Homopolymers

Polymer	Solvent	Temp (°C)	$K \times 10^2$ (ml/g)	a	MW range ($M \times 10^{-3}$)	Remark	Ref
Polystyrene	THF	20	1.55	0.692			1
		23	0.682	0.766	50–1000	A	2
		25	1.11	0.725		B	3
			1.60	0.706	> 3		4
			1.17	0.725		B	5
			1.17	0.717		B	6
			1.25	0.707			7
			1.68	0.69			8
			1.41	0.70			9
			1.25	0.717			10
			1.42	0.70			11
			1.16	0.724			12
			1.28	0.70			13
			0.609	0.768			14
		30	1.76	0.679			15
			0.86	0.74		B	16
			1.05	0.731	$36-10^4$		17
			1.57	0.698			18
			1.05	0.731		B	19
			1.62	0.701		B	20
			1.112	0.723			21
			1.28	0.712	$> 10^4$	C	22
			17.1	0.428	$< 10^4$		
		31.15	1.25	0.717	4–1800		23
		35	1.258	0.715			24
			1.580	0.704			25
			1.60	0.703			26
		38	1.247	0.720			27
		40	1.90	0.68			28
			1.31	0.719	19.1–1110	B	29
			1.51	0.706	5.11–1030	A	
	benzene	25	1.13	0.73			30
		30	1.623	0.695	5–400		31
	toluene	25	1.05	0.73			32
			0.662	0.76			33
			11.4	0.470	< 9000		34
			1.06	0.728	> 9000		

Polymer	Solvent	Temp (°C)	$K \times 10^2$ (ml/g)	a	MW range ($M \times 10^{-3}$)	Re-mark	Ref
		30	1.786	0.680	5–2000		31
			1.20	0.714	17–10^4		17
			9.86	0.499	0.56–17		
			1.57	0.695		B	35
		60	0.797	0.75			36
		68	1.528	0.691			15
	chloroform	25	0.716	0.76	100–2800		37
		30	1.36	0.708	24–10^4		17
			11.1	0.502	0.16–24		
			0.49	0.796			38
	p-dioxane	30	2.241	0.655	5–100		31
	MC	25	1.533	0.695			39
		30	1.157	0.718			40
	MC/DCA	30	1.4	0.70		D	41
	cyclohexane	34.5	8.3	0.50		B	35
		44.5	2.30	0.608		B	
	cyclohexanone	40	3.02	0.643		E	42
	MEK	30	1.79	0.642		B	35
	DMF	20	2.40	0.63			43
		30	2.606	0.612	5–400		31
	DMF + 0.01 M LiBr	60	1.30	0.662			44
	DMF	135	2.80	0.606			45
	OCP	25	14.3	0.488			46
		90	1.2	0.70			28
	TCB	127	1.21	0.71			47
		130	0.895	0.727		A	2
		135	1.21	0.707			48
			1.72	0.67			49
		140	1.90	0.655			50
			1.34	0.70			51
		145	4.14	0.610			52
			3.53	0.617			
			1.75	0.67			53
	ODCB	87	1.0	0.73		B	54
		135	1.156	0.715			22
			1.38	0.70	2–900		55
			4.87	0.606			56
			1.51	0.693			57
		138	1.38	0.7		F	58
	m-cresol	130	0.846	0.715		B	59
		135	2.02	0.65	4–200		28
	cis-decalin	25	4.0	0.574	20–24000	B	60
	HFIP/ toluene (2:8)	50	1.795	0.657			61
	Phenol/TCB (1:1)	135	1.303	0.69			62
	CNP	208	1.86	0.657			63
PMS	THF	25	4.2	0.608			11
	TCB	135	1.61	0.672			48

Polymer	Solvent	Temp (°C)	$K \times 10^2$ (ml/g)	a	MW range ($M \times 10^{-3}$)	Remark	Ref
PVC	THF	25	1.63	0.766	20–170		4
			1.50	0.77	20–100	B	6
			7.2	0.61	27–100		8
			4.48	0.70			14
			1.6	0.77		G	64
			4.76	0.673			65
		30	1.60	0.77	100–1000	B	5
		35	1.556	0.690			24
			1.8	0.76		B	66
	TCB	110	1.61	0.762			67
PVAc	THF	25	3.50	0.63	10–1000	B	5
			0.942	0.737			12
			0.51	0.791			68
			1.6	0.70			69
		30	1.49	0.698			70
		35	1.56	0.708			71
	TCB	135	1.37	0.687			70
PMMA	THF	23	0.93	0.69			8
		25	1.28	0.69	150–1200		9
			0.958	0.695			13
			0.787	0.75			65
			1.48	0.677			72
			1.99	0.660			73
		35	1.22	0.69			26
	chloroform	25	1.47	0.714			73
		30	0.43	0.80			38
	acetone	25	3.66	0.578			73
	MEK	25	0.71	0.72			38
	benzene	25	0.55	0.76			30
			4.38	0.597			73
		30	0.52	0.76			74
	toluene	25	1.91	0.68			33
			2.08	0.64	> 7800		34
	cyclohexanone	40	3.66	0.603		E	42
	DMF	25	0.404	0.787		A	75
			2.5	0.625		B	
	TFE	30	0.281	0.88			76
(100% isotactic)	THF	25	1.665	0.660			77
(92% iso)			1.690	0.659			
(83% iso)			1.719	0.658			
(35% iso)			0.984	0.692			
(23% iso)			0.9577	0.695			
(15% iso)			0.8892	0.694			
PEMA	THF	25	1.549	0.679			14
			2.25	0.66			78
PBMA	THF	25	0.503	0.758			14
	MEK	25	0.613	0.7258	10^4–2×10^4		79
			0.970	0.68	10^5–7×10^5		80

Polymer	Solvent	Temp (°C)	$K \times 10^2$ (ml/g)	a	MW range ($M \times 10^{-3}$)	Remark	Ref
POMA	THF	25	5.556	0.560			14
PLMA	THF	25	0.73	0.69			81
PSMA	THF	30	0.90	0.67	B		82
PMA	THF	25	0.388	0.82			83
	toluene	35	2.1	0.60			84
PtBA	THF	25	0.33	0.80	H		85
Polyethylene	ODCB	135	4.77	0.70	6–700		56
			4.90	0.74			86
		138	5.06	0.7		F	58
	TCB	135	5.50	0.704			87
			4.34	0.724			48
			9.54	0.64			49
			5.3	0.7		F	88
HDPE	ODCB	135	5.05	0.693	10–1000		57
	TCB	127	3.58	0.74			47
		130	3.92	0.725			89
		135	9.54	0.64			90
			5.26	0.70			88
			5.23	0.70			91
			7.1	0.67			92
			5.63	0.70			93
		140	3.95	0.726			50
			3.23	0.735			94
		145	1.73	0.784			52
	decalin	135	4.17	0.71			95
			3.9	0.74			96
			4.6	0.73			91
		138	6.2	0.7		F	97
	decane	120	6.98	0.65			95
LDPE	ODCB	135	5.06	0.70	0.2–200		99
LLDPE	TCB	145	5.91	0.69			53
Hydrogenated PB	TCB	135	2.7	0.746	2.4–40		58
PP	TCB	135	1.37	0.75			48
			1.76	0.73			91
		145	1.56	0.76			53
	ODCB	135	1.0	0.78			86
			2.42	0.707		B	98
			1.10	0.80			99
	decalin	135	1.10	0.80			86, 91
Polyisobutylene	THF	25	5	0.60		H	100
		40	6.47	0.593	0.73–67.6	A	29
			5.79	0.593	0.82–80.4	B	
Polycarbonate	THF	25	3.99	0.77		I	28
			4.9	0.67	7–270		101
			4.12	0.69	10–300	J	102
		35	3.18	0.680		K	25
PB	toluene	30	5.19	0.679	0.5–10^4		17
			2.888	0.734			21
		68	2.103	0.753			15

Polymer	Solvent	Temp (°C)	$K \times 10^2$ (ml/g)	a	MW range ($M \times 10^{-3}$)	Re- mark	Ref
	THF	25	1.012	0.621			27
			18.4	0.58			65
			1.6	0.776			67
		30	4.57	0.693	$0.5-10^4$		17
			2.56	0.74		B	82
		40	3.96	0.697		L	103
	chloroform	30	4.51	0.704	$0.19-10^4$		17
		40	5.78	0.67	$10-100$		28
(*cis/trans* = 0.8)		30	76.0	0.44			
(20% *cis*-20% vinyl)		25		2.36	0.75	3-6	
(8% vinyl)			4.57	0.693	$80-1100$		10
(28% vinyl)			4.51	0.693	$20-300$		
(52% vinyl)			4.28	0.693	$20-200$		
(73% vinyl)			4.03	0.693	$20-200$		
SBR(25% styrene)	THF	25	4.10	0.693	$2.4-40$		
		40	3.18	0.70	$70-1000$		28
Natural rubber							
(poly-*cis*-isoprene)	THF	25	1.09	0.79	$10-1000$	G	
Polyisoprene	THF	25	1.77	0.753	$40-500$		10
Butyl rubber	THF	25	0.85	0.75	$4-4000$		28
PDSX	toluene	25	0.828	0.72			38
		60	0.977	0.725			36
	chloroform	30	0.54	0.77			38
	cyclohexane	35	0.63	0.77			104
	ODCB	87	8.19	0.50	$20-800$		54
		138	3.83	0.57	$25-300$		55
	DMAC	25	1.3	0.69		M	105
Nylon 6	*m*-cresol	25	32	0.62	$0.5-5$		28
	OCP	90	6.2	0.64			
	HFB	25	4.2	0.76			106
	TFE	25	4.58	0.742	$20-100$		107
			5.36	0.75	$13-100$		108
Nylon 66	*m*-cresol	25	3.53	0.79	$0.15-50$		28
		130	4.01	1.00	$8-24$	B	59
	HFIP		19.8	0.63			109
Nylon 610	*m*-cresol	25	1.35	0.96	$8-24$		28
Nylon 11	*m*-cresol	30	9.1	0.69			110
Nylon 12	*m*-cresol	25	4.6	0.75	$10-125$		28
	HFIP/ toluene (2:8)	50	6.163	0.6585			61
Trogamid T	DMF	22	2.737	0.706			111
TFA-nylon 6	MC	25	3.355	0.64		N	112
	THF		1.66	0.70		O	113
TFA-nylon 12	MC	30	0.738	0.81		P	114
TFA-Trogamid	THF	25	0.87	0.73		O	113
PET	*m*-cresol	25	0.077	0.95			28
		135	1.75	0.81	$27-32$		
			2.0	0.90	$0.45-0.8$		
	OCP	23	6.31	0.658			115
	PTCE	23	7.44	0.648			

Polymer	Solvent	Temp (°C)	$K \times 10^2$ (ml/g)	a	MW range ($M \times 10^{-3}$)	Re- mark	Ref
	HFIP	23	5.20	0.695			
	PFP	23	3.85	0.723			
	HFIP/PFP (1:1)	23	4.50	0.705			
		25	6.31	0.658			46
	MC/DCAA	30	4.5	0.68		Q	41
	MC/HFIP (7:3)	25	4.034	0.691		R	116
LCA polyester	PFP	50	0.9205	1.05			117
PTHF	cyclohexanone	40	4.16	0.676		E	42
Polyoxymethylene	DMF	135	0.932	0.79			45
PVDBC	THF	25	9.1	0.827			118
PVB	THF	25	1.4	0.80		H	119
POA	THF	25	0.842	0.202	> 12.5	G	120
			0.961	0.674	< 12.5	G	
PDEPP	THF	25	2.5	0.65		S	121
PDPPP, PAPP	THF	30	1.19	0.649		B, T	20
PVK	THF	25	0.127	0.80		U	122
PPS	CNP	210	0.891	0.747			63
PEEK	sulfuric acid	25	0.3849	0.94		B	62
	phenol/TCB (1:1)	115	7.588	0.67		B	62

Abbreviations. THF – tetrahydrofuran; MC – methylene chloride; MEK – methyl ethyl ketone; DMF – N,N-dimethylformamide; OCP – o-chlorophenol; TCB – 1, 2,4-trichlorobenzene; ODCB – o-dichlorobenzene; HFIP – hexafluoroisopropanol; CNP – 1-chloronaphthalene; TFE – trifluoroethanol; DMAC – N,N-dimethylacetamide; HFB – hexafluorobutanol; PTCE – phenol/tetrachloroethane (6:4); PFP – pentafluorophenol; PMS – poly(methyl styrene); PVC – poly(vinyl chloride); PVAc – poly(vinyl acetate); PMMA – poly(methyl methacrylate); PEMA – poly(ethyl methacrylate); PBMA – poly(butyl methacrylate); POMA – poly(octyl methacrylate); PLMA – poly-(n-lauryl methacrylate); PSMA poly(stearyl methacrylate); PMA – poly(methyl acrylate); PtBA – poly-($tert$-butyl acrylate); HDPE – high density polyethylene (PE); LDPE – low density PE; LLDPE – linear LDPE; PP – polypropylene; PB – polybutadiene; SBR – styrene-butadiene rubber; PDSX – polydimethylsiloxane; Trogamid T – poly(trimethylhexamethylene terephthalamide); PET – poly(ethylene terephthalate); LCA polyester-liquid crystalline aromatic polyester; PTHF – poly(tetrahydrofuran); PVDBC – poly(N-vinyl-3,6-dibromocarbazole); PVB – poly(vinyl butyral); POA – poly[N-(n-octadecyl) maleimide]; PDEPP – poly(diethoxyphospazene); PDPPP – poly(diphenyl phosphazene); PAPP – poly-(aryloxy phosphazen); PVK – poly(N-vinyl-3,6-dibromocarbazole); PPS – poly(phenylene sulfide); PEEK – poly(arylether ether ketone).

Remarks. A – $[\eta] = KM_n^a$; B – $[\eta] = KM_w^a$; C – narrow MW distribution; D – MC/dichloroacetic acid (8:2) + 0.01 M tetrabutylammonium acetate; E – 0.01 M tetrabutylammonium nitrate included; F – $[\eta] = KM_v^a$; G – use K and a from [4] for PS; H – use K and a from [6] for PS; I – 4,4'-dihydroxy-2,2-propane carbonate; J – use K and a from [9] for PS; K – bisphenol A/ diethylene glycol (50:50); L – 35% cis, 55% trans, 10% 1,2; M – poly(methyl) (pyridine-3-yl) siloxane; N – use K and a from [37] for PS; O – use $K = 1.31 \times 10^{-2}$ and $a = 0.714$ for PS; P – use K and a from [38] for PS; Q – MC/dichloroacetic acid + 0.01 M tetrabutylammonium acetate; R – use $K = 7.998 \times 10^{-2}$ and $a = 0.54$ for PS; S – 0.1% tetra(n-butylammonium) bromide included; T – 0.1 M LiBr included; U – use K and a from [3] for PS.

References

1. ECHARRI J, IRUIN JJ, GUZMAN GM, AMSORENA J (1979) Makromol Chem 180:2749
2. BONI KA, SLIEMERS FA, STICKNEY PB (1968) J Polym Sci Part A 26:1579
3. SPATORICO AL, COULTER B (1973) J Polym Sci Polym Phys Ed 11:1139
4. PROVDER T, ROSEN EM (1970) Sep Sci 5:437
5. GOEDHAR D, OPSHOOR A (1970) J Polym Sci Part A-2, 8:1227
6. KOLÍNSKY M, JANČA J (1974) J Polym Sci Polym Chem Ed 12:1181
7. MORRIS MC (1971) J Chromatogr 55:203
8. GRUBISIC Z, REMPP P, BENOIT H (1967) J Polym Sci Part B 5:753
9. RUDIN A, HOEGY HW (1972) J Polym Sci Part A-1, 10:217
10. KRAUS C, STACY CJ (1972) J Polym Sci Part A-2, 10:657
11. SCHRODER UKO, EBERT KH (1987) Makromol Chem 188:1415
12. COLEMAN TA, DAWKINS JV (1986) J Liq Chromatogr 9:1191
13. JENKINS R, PORTER RS (1980) J Polym Sci Polym Lett Ed 18:743
14. SAMAY G, KUBÍN M, PODEŠVA J (1978) Angew Makromol Chem 72:185
15. AMBLER MR (1973) J Polym Sci Polym Chem Ed 11:191
16. WILLIAMS RC, SCHMIT JA (1971) J Polym Sci Part B 9:413
17. AMBLER MR (1980) J Appl Polym Sci 25:901
18. SUDDABY KG, SANAYEI RA, O'DRISCOLL KF, RUDIN A (1993) Makromol Chem 194:1965
19. AMBLER MR, McINTYRE D (1977) J Appl Polym Sci 21:2269
20. DEJAEGER R, LECACHEUX D, POTIN PH (1990) J Appl Polym Sci 39:1793
21. SMITH WV (1974) J Appl Polym Sci 18:3685
22. HANEY MA, ARMONAS JE, ROSEN L (1987) In: PROVDER T (ed) ACS Symposium Series No. 352, ACS, Washington, p 119
23. VAN DIJK JAPP, SMIT JAM, KOHN FE, FEIJEN J (1983) J Polym Sci Polym Chem Ed 21:197
24. PERKINS G, HAEHN (1990) J Vinyl Technol 12:2
25. MAHABADI HKH (1989) International GPC Symposium, Newton, Mass, 434
26. MAHABADI HKH (1985) J Appl Polym Sci 30:1535
27. ITO K, SAITO T, AOYAMA T (1987) Polymer 28:1589
28. EVANS JM (1973) Polym Eng Sci 13:401
29. XIE J (1994) Polymer 35:2385
30. TSITSILIANIS C, MITSIANI G, DONDOS A (1989) J Polym Sci Part B Polym Phys 27:763
31. ZINBO M, PARSONO JL (1971) J Chromatogr 55:55
32. WINTERMANTEL M, ANTONIETTI M, SCHMIDT M (1993) J Appl Polym Sci, Appl Polym Symp 52:91
33. PRICE GJ, MOORE JW, GUILLET JE (1989) J Polym Sci Part A Polym Chem 27:2925
34. SANAYEI RA, O'DRISCOLL KF (1991) J Macromol Sci Chem A 28:987
35. VARMA BK, FUJITA Y, TAKAHASHI M, NOSE T (1984) J Polym Sci Polym Phys Ed 22:1781
36. MANDIK L, FOKSOVÁ A, FOLTYN J (1979) J Appl Polym Sci 24:395
37. OTH J, DESREUX V (1954) Bull Soc Chim Belges 63:285
38. DAWKINS JV (1968) J Macromol Sci Phys B 2:623
39. BIAGINI E, COSTA G, GATTIGLIA E, IMPERATO A, RUSSO S (1987) Polymer 28:114
40. OGAWA T, SAKAI M (1988) J Polym Sci Part A Polym Chem 26:3141
41. MOUREY TH, BRYAN TG, GREENER J (1993) J Chromatogr A 657:377
42. MOUREY TH, MAILLER SM, FERRAR WT, MOLAIRE TR (1989) Macromolecules 22:4286
43. KRANZ D, POHL HU, BAUMANN H (1972) Angew Makromol Chem 26:67
44. AZUMA C, DIAS ML, MANO EB (1986) Macromol Chem Macromol Symp 2:169
45. OGAWA T (1990) J Liq Chromatogr 13:51
46. MARTIN L, Lavine M, BALKE ST (1992) J Liq Chromatogr 15:1817
47. SCHEINERT W (1977) Angew Makromol Chem 63:117
48. COLL H, GILDING DK (1970) J Polym Sci Part A-2 6:1579
49. WILLIAMSON GR, CERVENKA A (1972) Eur Poly J 8:1009
50. BARLOW A, WILD L, RANGANATH R (1977) J Appl Polym Sci 21:3319

51. KAŠPARKOVA V, OMMUNDSEN E (1993) Polymer 34:1765
52. LEW R, CHEUNG P, BALKE ST, MOUREY TH (1993) J Appl Polym Sci 47:1685
53. GRINSHPUN V, RUDIN A (1985) Makromol Chem Rapid Commun 6:219
54. DAWKINS JV, HEMMING M (1972) Makromol Chem 155:75
55. DAWKINS JV, MADDOCK JW, COUPE DJ (1970) J Polym Sci Part A-2 8:1803
56. TROTH HG (1968) 5th International Seminar for GPC
57. POLLOCK DJ, KRATZ RF (1968) 6th International Seminar for GPC
58. DAWKINS JV, MADDOCK JW (1971) Eur Polym J 7:1537
59. DUDLEY MA (1972) J Appl Polym Sci 16:493
60. KULICKE WM, PRESCHER M (1984) Makromol Chem 185:2619
61. OGAWA T, SAKAI M, ISHITOBI W (1985) J Polym Sci Polym Chem Ed 24:109
62. DEVAUX J, DELIMOY D, DAOUST D, LEGRAS R, MERCIER JP (1985) Polymer 26:1994
63. STACY CJ (1986) J Appl Polym Sci 34:3959
64. LEDERER K, AMTMANN I, VIJAYAKUMAR S, BILLIANI J (1990) J Liq Chromatogr 13:1849
65. CHAPLIN RP, HAKEN JK, PADDON JJ (1979) J Chromatogr 171:55
66. HATTORI S, ENDOH H, NAKAHARA H, KAMATA T, HAMASHIMA M (1978) Polym J 10:173
67. PANG S, RUDIN A (1993) J Appl Polym Sci 49:1189
68. PARK WS, GRAESSLEY WW (1977) J Polym Sci Polym Phys Ed 15:71
69. CANE F, CAPACCIOLI T (1978) Eur Poly J 14:185
70. LECACHEUX D, LESEC J, QUIVORON C, PRECHNER R, PANARAS R, BENOIT H (1984) J Appl
    Polym Sci 29:1569
71. ATKINSON CML, DIETR R (1979) Eur Polym J 15:21
72. CHEE KK (1985) J Appl Polym Sci 30:1323
73. DOBBIN CJB, RUDIN A, TCHIR MF (1982) J Appl Polym Sci 27:1081
74. COHN-GINSBERG E, FOX TG, MASON HF (1962) Polymer 3:97
75. KOSSLER I, NETOPILIK M, SCHULZ G, GNAUCK R (1982) Polym Bull 7:597 (1982)
76. VEITH CA, COHEN RE (1989) Polymer 30:942
77. JENKINS R, PORTER RS (1982) Polymer 23:105
78. GRUBISIC-GALLOT Z, LINGELSER JP, GALLOT Y (1990) Polym Bull 23:389
79. SIMIONESCU CI, SIMIONESCU BC, NEAMTU I, IOAN S (1987) Polymer 28:165
80. FEDORS RF (1979) Polymer 20:226
81. MAHABADI HKH, O'DRISCOLL KH (1977) J Appl Polym Sci 21:1283
82. ZHONGDE X, MINGSHI S, HADJICHRISTIDIS N, FETTERS LJ (1981) Macromolecules
    14:1591
83. SZESZTAY M, TUEDOES F (1981) Poly Bull 5:429
84. KRAUNAKARAN K, SANTAPPA M (1968) J Polym Sci Part A-2, 6:713
85. MRKVIČKOVA L, DANHELKA J (1990) J Appl Polym Sci 41:1929
86. OGAWA T, INABA T (1977) J Appl Polym Sci 21:2979
87. LEHTINEN A, VAINIKKA R (1989) International GPC Symposium, Newton, MA, 612
88. PEYROUSET A, PRECHNER R, PANARIS R, BENOIT H (1975) J Appl Polym Sci 19:1363
89. WAGNER HL, HOEVE CAJ (1973) J Polym Sci Polym Phys Ed 11:1189
90. WILLIAMSON GR, CERVENKA A (1972) Eur Polym J 8:1009
91. CROUZET P, MARTENS A, MANGIN P (1971) J Chromatogr Sci 9:525
92. COTE JA, SHIDA M (1971) J Polym Sci Part A-2, 9:421
93. HERT M, STRAZIELLE C (1983) Makromol Chem 184:135
94. WILD L, RANGANATH R, RYLE T (1971) J Polym Sci Part A-2, 9:2137
95. BOHM LL, LANVER U, LECHNER MD (1983) Makromol Chem 184:585
96. DE LA CUESTA MO, BILLMEYER FW Jr (1963) J Polym Sci A-1, 1:1721
97. CHIANG R (1965) J Phys Chem 69:1645
98. ATKINSON CML, DIETZ R (1976) Makromol Chem 177:213
99. OGAWA T, SUZUKI Y, INABA T (1972) J Polym Sci Part A-1, 10:737
100. MRKVIČKOVA L, LOPOUR P, POKORNY S, JANČA J (1980) Angew Makromol Chem 90:217
101. MOORE WR, UDDIN M (1969) Eur Polym J 5:185
102. BAILLY C, DAOUST D, LEGRAS R, MERCIER JP, STRAZIELLE C, LAPP A (1986) Polymer
     27:1410

103. BARLOW A, WILD L, ROBERTS T (1971) J Chromatogr 55:155
104. DAWKINS JV, HEMMING M (1975) Makromol Chem 176:1777
105. VILENCHIK LZ, RUBINSTAJN S, ZELDIN M, FIRE WK (1991) J Polym Sci Part B Polym Phys 29:1137
106. COSTA G, RUSSIO S (1982) J Macromol Sci Chem Ed A18:299
107. AHARONI SM, CILURSO FG, HANRAHALN JM (1985) J Appl Polym Sci 30:2505
108. MATTIUSSI A, GECHELE GB, FRANCESCONI R (1969) J Polym Sci A-2, 7:411
109 SAMANTA SR (1992) J Appl Polym Sci 45:1635
110. TERAN CC, MACCHI EM, FIGINI RV (1987) J Appl Polym Sci 34:2433
111. HEROLD J, MEYERHOFF G (1980) Makromol Chem 181:2625
112. BIAGINI E, COSTA G, GATTIGLIA E, IMPERAT A, RUSSO S (1987) Polymer 28:114
113. WEISSKOPF K (1985) Polymer 26:1187
114. OGAWA T, SAKAI M (1988) J Polym Sci Part A Polym Chem 26:3141
115. BERKOWITZ S (1984) J Appl Polym Sci 29:4353
116. OVERTON JR, BROWING HL Jr (1984) In: PROVDER T (ed) ACS Symposium Series No. 245, ACS, Washington, p 219
117. KINUGAWA A (1991) J Liq Chromatogr 14:1315
118. HORTA A, SAÍZ E, BARRALES-RIENDA JM, GÓMEZ PAG (1986) Polymer 27:139
119. MRKVIČKOVA L, DANHELKA J, POKORY S (1984) J Appl Polym Sci 29:803
120. BARRALES-RIENDA JM, GALIECA CR, HORTA A (1983) Macromolecules 16:932
121. TARAZONA MP, BRAVO J, RODRIGO MM, SAIZ E (1991) Polym Bull 26:465
122. BARRALES-RIENDA JM, GOMEZ PAG, HORTA A, SAIZ E (1985) Macromolecules 18:2572

Appendix II

Mark–Houwink Parameters for Polar Polymers

Polymer	Solvent	Temp (°C)	$K \times 10^2$ (ml/g)	a	MW range ($M \times 10^{-3}$)	Remark	Ref
PAN	DMF	20	3.25	0.725			1
		25	2.43	0.75			2
		60	0.146	0.897		A	3
			2.12	0.75		A	4
PVP	DMF	60	1.85	0.646		A	3
	MeOH/H$_2$O (1:1)	25	0.851	0.72		B	5
P2VP	THF	25	2.23	0.66			6
		35	3.02	0.61			7
	TFA + NaNO$_3$	35	0.575	0.922		C	
			1.62	0.745		D	
			2.57	0.651		E	
	pyridine	25	0.445	0.78		F	8
	NMP	80	0.88	0.73		G	
	DMF	50	1.23	0.69		H	
	sodium formate	25	0.25	0.93		I	9
			0.50	0.83		J	
			0.17	0.95		K	
	see Remark	25	2.5	0.68		L	10
Polyacrylamide	water	25	6.80	0.66			11
			1.0	0.755	40–9000	M	12
		30	0.631	0.80			13
	0.1 M Na$_2$SO$_4$	25	1.94	0.70			14
	0.5 M NaCl	25	1.21	0.746		pH 9, M	15
	1 M NaCl		1.84	0.72		pH 9, M	
NaPA	0.3 N NaCl	25	1.69	0.75			16
Phenol (RN)	acetone	25	4.7	1.90	1–8, M$_n$		17
		30	63.1	0.28	0.37–28, M$_n$		18
	THF	30	73	0.28	0.54–4.7, M$_n$		
	DMF	25	18.0	0.51	1–8, M$_n$		17
	1N NaOH		16.6	0.48			
(HO)	acetone	30	8.13	0.5	0.69–2.6, M$_n$		18
PVPDMAEMA	LiNO$_3$	25	1.42	0.67		N	19
PTAC 50	1 M NaCl	25	5.64	0.90		O	20
PTAC 100	1 M NaCl	25	1.55	0.97		O	20
PTMAC 50	1 M NaCl	25	185	0.66		O	20
PTMAC 100	1 M NaCl	25	59.8	0.72		O	20
PMVEMA	tris-buffer	35	3.17	0.70		P	21
PDMDAAC	0.1 M NaCl	30	3.5	0.62		O	22

Polymer	Solvent	Temp (°C)	$K \times 10^2$ (ml/g)	a	MW range ($M \times 10^{-3}$)	Re-mark	Ref
Poly(N-vinylacetamide)							
	H_2O	30	16	0.52		Q	22
Polyamic acid	DMAc	30	2.01	0.79		R	23
PEG, PEO	H_2O	25	1.25	0.78			24
		30	1.25	0.78			13
	$LiNO_3$	25	2.8	0.72		N	19
		35	3.48	0.70		P	21
PEG	0.1 N $NaNO_3$	35	10.1	0.58			25
PEO			3.47	0.70			
	0.42 N NaOH	25	2.5	0.66			26
	p-dioxane	25	3.50	0.71	0.06–19		27
	DMF	25	2.40	0.73	1–30		
		60	5.0	0.653		A	3
			5.50	0.643		A	4
	$MeOH/H_2O$ (1:1)	25	2.21	0.75		B	5
PEG	$MeCN/H_2O$ (2:8)	35	87.3	0.60			25
PEO			2.45	0.73			
	$MeOH/H_2O$ (2:8)	25	3.4	0.75			28
PPG	benzene	30	3.241	0.686	1–4		27
	toluene	30	0.447	0.920	0.8–4		
	p-dioxane	30	2.247	0.715	1–4		
	DMF	30	5.284	0.598	0.8–4		
PVA	H_2O	25	14.0	0.60	10–70		29
	0.1 N $NaNO_3$	35	13.3	0.56	13–250		30
Cellulose nitrate	THF	25	25.0	1.00	95–2300		31
			82	1.0	DP < 1	S	32, 33
			446	0.76	DP > 1	S	
			6.44	0.73		T	34
Nitrocellulose	THF	25	3.21	0.83			35
Methyl cellulose	H_2O	25	10.7	0.61		O	36
Dextran	H_2O	25	9.78	0.5			24
	0.1 M NaCl	23	15.7	0.451		U	37
		25	13.6	0.47			38
	0.1 M $NaNO_3$	25	37	0.40		V	39
	0.2 M $NaNO_3$		48.5	0.39			40
	0.2 M $NaNO_3$ + 0.025 M KH_2PO_4	30	47.6	0.34			41
	0.42 N NaOH	25	12.0	0.48			26
	NaOH (0.005 N)	25	18.0	0.438			42
	(0.0185 N)		18.5	0.431			
	(0.097 N)		15.8	0.451			
	(0.287 N)		12.4	0.474			
	(0.501 N)		13.2	0.478			
	(0.671 N)		14.7	0.465			
	(0.877 N)		15.1	0.464			
Branched dextran	$MeOH/H_2O$		103	0.25			43
Pullulan	0.1 M NaCl	25	1.79	0.67			38
	0.2 M $NaNO_3$		4.22	0.64			40
	0.1 M $NaNO_3$ + 0.025 M KH_2PO_4	30	1.70	0.72			41

Polymer	Solvent	Temp (°C)	$K \times 10^2$ (ml/g)	a	MW range $(M \times 10^{-3})$	Re-mark	Ref
	NaAc buffer	25	1.79	0.67			44
	H_2O	25	2.23	0.67		Q	45
			2.36	0.66			46
Polysaccharide	0.1 N NaNO$_3$	35	2.14	0.66			25
	MeCN/H$_2$O (2:8)	35	3.72	0.59			
	MeOH/H$_2$O (2:8)	25	4.4	0.6			28
	H_2O	25	6.9	0.6			
Xanthan	0.1 M NaNO$_3$	25	0.024	1.10		V	39
Xylan	0.5 M NaOH	25	2.67	0.73			47
NaPSS	phosphate	25	0.60	0.74		Q, W	45
		25	0.50	0.80		Q, X	45
	NaCl (0.005 N)	25	0.057	1.061			42
	(0.05 N)		0.106	0.935			
	(0.20 N)		0.209	0.845			
	(0.50 N)		0.285	0.794			
	(1.0 N)		0.416	0.741			
	0.1 M NaCl	60	0.24	0.80			48
	NaOH (0.005 N)	25	0.038	1.079			42
	(0.0185 N)		0.146	0.935			
	(0.097 N)		0.458	0.805			
	(0.287 N)		0.281	0.815			
	(0.501 N)		0.198	0.828			
	(0.671 N)		0.174	0.830			
	(0.877 N)		0.110	0.851			
	0.42 N NaOH	25	0.2	0.82			26
	see Remark		1.09	0.74		Y	10
Chitosan	see Remark	25	19.9	0.59		O, Z	49
HA	0.15 M NaCl	25	6.54	1.16			50
	see Remark		3.6	0.78		Z2	51
Ch4-S	0.2 M NaCl	25	17	1.01			50
	0.02 M NaCl		3.1	0.74			52
P(DL)LA	THF	31.15	5.49	0.639			53
Proteins	guanidine		3.16	0.66		Z3	54
	see Remark	25	0.0174	1.09		Z4	55

Abbreviations. DMF – *N*,*N*-dimethylformamide; THF – tetrahydrofuran; TFA – trifluoro-acetic acid; NaAc – sodium acetate; phosphate – phosphate buffer; NMP – *N*-methylpyrroli-done; PAN – poly(acrylonitrile); PVP – poly(vinyl pyrrolidone); P2VP – poly(2-vinyl pyridine); NaPA – sodium polyacrylate; phenol (RN) – phenol-formaldehyde resin (random novolak resin); Phenol(HO) – phenol-formaldehyde resin (high-ortho resin); PEG – poly-(ethylene gylcol); PEO – poly(ethylene oxide); PPG – poly(propylene glycol); PVA – poly-(vinyl alcohol); NaPSS – sodium poly(styrene sulfonate); PVPDMAEMA – poly(vinyl-pyrrolidone-co-dimethylaminoethyl methacrylate); PTAC 50 – poly(acrylamide-co-*N*,*N*,*N*-trimethylammonium ethyl acrylate chloride); PTAC 100 – poly(*N*,*N*,*N*-trimethylammonium ethyl acrylate chloride); PTMAC 50 – poly(acrylamide-co-*N*,*N*,*N*-trimethylammonium ethyl methacrylate chloride); PTMAC 100 – poly(*N*,*N*,*N*-trimethylammonium ethyl methacrylate chloride); PMVEMA – poly(methyl vinyl ether-co-maleic anhydride); PDMDAAC – poly-(dimethyldiallylammonium chloride); HA – hyaluronic acid; Ch4-S – chondroitin 4-sulfate; P(DL)LA – poly(DL-lactic acid).

Remarks. A – 0.01 M LiBr included; B – 0.01 M LiNO$_3$ included; C – 0.1% TFA + 0.01 N NaNO$_3$; D – 0.1% TFA + 0.20 N NaNO$_3$; E – 9.1% TFA + 0.50 N NaNO$_3$; F – use K = 0.379 × 10^{-2} and a = 0.81 for PS; G – use K = 1.8 × 10^{-2} and a = 0.68 for PS; H – use K = 2.88 × 10^{-2} and a = 0.60 for PS; I – 0.1 M sodium formate at pH 3.50; J – 0.5 M sodium formate at pH 3.50; K – 0.1 M sodium formate + 20% methanol at pH 3.55; L – 80%(0.5 M Na$_2$SO$_4$ + 0.5 M MeCOOH) + 20% MeCN; M – [η] = KM_v^a; N – 0.5 M LiNO$_3$ (pH 7.0); O – [η] = KM_w^a; P – 0.1 M tris (hydroxymethyl amino methane) with 0.2 M LiNO$_3$ adjusted to pH 9.0 with HNO$_3$; Q – [η] = KM_n^a; R – 0.03 M LiBr + 0.03 M H$_3$PO$_4$ + 1 vol% THF in dimethyl acetamide, use K = 1.21 × 10^{-2} and a = 0.69 for PS; S – [η] = K(DP)^a, DP – degree of polymerization; T – 0.01 M acetic acid included; U – 0.02 M disodium hydrogen phosphate at pH 7 included; V – 0.2 g/l NaN$_3$ + 0.1% ethylene glycol included; W – 0.51 M phosphate buffer (pH 8.0); X – 0.05 M phosphate buffer (pH 8.0); Y – 80% of 0.03M Na$_2$SO$_4$ + 20% MeCN; Z – 0.5 M acetic acid + 0.5 M sodium acetate; Z2 – 0.1 M phosphate + 0.1 M NaCl; Z3 – 6 M guanidinium hydrochloride; Z4 – 0.1 M sodium phosphate + 2% sodium dodecyl sulfate + 0.02 M dithiothreitol (pH 7.0), proteins (bovine serum albumin, chymotrypsinogen, lysozyme chloride, and cytochrome C).

References

1. KRANZ D, POHL HU, BAUMANN H (1972) Angew Makromol Chem 26:67
2. KENYON AS, MOTTUS EH (1974) Appl Polym Symp 25:57
3. MORI S (1983) Anal Chem 55:2414
4. AZUMA C, DIAS ML, MANO EB (1986) Macromol Chem Macromol Symp 2:169
5. SENAK L, WU CS, MALAWER EG (1987) J Liq Chromatogr 10:1127
6. HUGELIN C, DONDOS A (1969) Macromol Chem 126:207
7. NAGY DJ, TERWILLIGER DA, LAWREY BD, TIEDGE WF (1989) International GPC Symposium, Newton, MA, 637
8. RAND WG, MUKHERJI AK (1982) J Chromatogr Sci 20:182
9. PATTON EV, WONNACOTT DM (1987) J Chromatogr 389:115
10. KÜHN A, FORSTER S, LÖSCH R, ROMMELFANGER M, ROSENAUER C, SCHMIDT M (1993) Makromol Chem Rapid Commun 14:433
11. ISHIGE T, HAMIELEC AE (1973) J Appl Polym Sci 17:1479
12. KULICKE WM, KNIEWSKE R, KLEIN J (1982) Prog Polym Sci 8:373
13. SAFIEDDINE AM, HESTER RD (1991) J Appl Polym Sci 43:1987
14. KULICKE WM, BÖSE N (1982) Polym Bull 7:205
15. MCCARTHY KJ, BURKHARDT CW, PARAZAK DP (1987) J Appl Polym Sci 33:1699
16. ROCHAS C, DOMARD A, RINAUDO M (1980) Eur Polym J 16:135
17. TOBIASON FL, CHANDLER C, SCHWARZ FE (1972) Makromolecules 5:321
18. KAMIDE K, MIYAKAWA Y (1978) Makromol Chem 179:359
19. WU CS, SENAK L (1990) J Liq Chromatogr 13:851
20. KULICKE WM, JACOBS A (1992) Makromol Chem Macromol Symp 61:59
21. WU CS, SENAK L, MALAWER EG (1989) J Liq Chromatogr 12:2919
22. DUBIN PL, LEVY IJ (1982) J Chromatogr 235:377
23. WALKER CC (1988) J Polym Sci Part A Polym Chem 26:1649
24. BAILEY FE, KUCERA JL, IMHOF LG (1958) J Polym Sci 32:517
25. NAGY DJ (1990) J Liq Chromatogr 13:677
26. CALLEC G, ANDERSON AW, TSAO GT (1984) J Polym Sci Polym Chem Ed 22:287
27. ZINBO M, PARSONO JL (1971) J Chromatogr 55:55
28. DAWKINS JV, GABBOTT NP, MONTENEGRO AMC (1990) J Appl Polym Sci Appl Polym Symp 45:103
29. DIEU HA (1954) J Polym Sci 12:417
30. NAGY DJ (1993) J Liq Chromatogr 16:3041
31. RUDIN A, HOEGY HW (1972) J Polym Sci Part A-1, 10:217
32. MARX-FIGINI M, SOUBELET O (1984) Polym Bull 11:281
33. SOUBELET O, PRESTA MA, MARX-FIGINI M (1990) Angew Makromol Chem 175:17

34. EREMEEVA TE, BYKOVA TO, GROMOV VS (1990) J Chromatogr 522:67
35. BROWN W, WILKSTROM R (1965) Eur Polym J 1:1
36. SAKAMOTO N (1987) Polymer 28:288
37. MAZSAROFF I, REGNIER FE (1988) J Chromatogr 442:15
38. KATO T, TOKUYAMA T, TAKAHASHI A (1983) J Chromatogr 256:61
39. TINLAND B, MAZET J, RINAUDO M (1988) Makromol Chem Rapid Commun 9:69
40. BAHARY WS, HOGAN MP, JILANI M, ARONSON MP (1995) In: PROVDER T, BARTH HG, URBAN MW (eds) Adv Chem Ser 247, ACS, Washington, p 151
41. BAHARY WS, HOGAN MP (1996) Int J Polym Anal Charact 2:121
42. BOSE A, ROLLINGS JE, CARUTHERS JM, OKOS MR, TSAO GT (1982) J Appl Polym Sci 27:795
43. CERNY L, MCTIERVAN J, STASIW D (1973) J Polym Sci Symp 42:1455
44. SOMMERMEYER K, CECH F, PFITZER E (1988) Chromatographia 25:167
45. DUBIN PL, TECKLENBURG MM (1985) Anal Chem 57:275
46. KAWAHARA K, OHTA K, MIYAMOTO H, NAKAMURA S (1984) Carbohydr Polym 4:335
47. EREMEEVA TE, BYKOVA TO (1993) J Chromatogr 639:159
48. MALFAIT T, SLOOTMAEKERS D, CAUWELAERT FV (1990) J Appl Polym Sci 39:571
49. YOMOTO C, MIYAZAKI T, OKADA S (1993) Colloid Polym Sci 271:76
50. HITTNER DM, COWNAN MK (1987) J Chromatogr 402:149
51. LAURENT TC, RYAN M, PIETRUSZKIEWICZ A (1960) Biochim Biophys Acta 42:476
52. MATHEWS MB, DORFMAN A (1953) Arch Biochem Biophys 42:41
53. VAN DIJK JAPP, SMIT JAM, KOHN FE, FEIJEN J (1983) J Polym Sci Polym Chem Ed 21:197
54. NAVE R, WEBER K, POTSCHKA M (1993) J Chromatogr A 654:229
55. CHIKAZUMI N, OHTA T (1991) J Liq Chromatogr 14:403

Appendix III

Mark–Houwink Parameters for Copolymers

Polymer	Solvent	Temp (°C)	$K \times 10^2$ (ml/g)	a	Remark	Ref
PVC-PVAc	THF	25	6.72	0.611	10–13% VAc	1
EPM	toluene	30	3.051	0.76	50% PP, A	2
EPDM	THF	35	27.4	0.54		3
EPR	ODCB	135	4.43	0.725		4
			2.42	0.707		5
	toluene	30	3.1	0.72	62% PE, B	6
E-NORB	xylene	90	10.0	0.535	50%	7
	DEB	120	4.93	0.589		
EVA	THF	20	9.7	0.62	27–29% EVA, C	8
P (S-MMA)	THF	ns	0.775	0.76	alternating	9
	MEK		1.15	0.69		
	toluene		1.09	0.73		
	n-chlorobutane		1.08	0.70		
	THF		0.641	0.76	1:1 block	
	MEK		1.466	0.67		
	toluene		1.318	0.69		
	n-chlorobutane		4.446	0.55		
	toluene	ns	1.14	0.70	statistical, 30% S	10
	n-chlorobutane		2.65	0.60		
	toluene		1.32	0.71	57% S	
	n-chlorobutane		2.49	0.63		
	toluene		0.832	0.75	71% S	
	n-chlorobutane		1.76	0.67		
P (S-MAH)	THF	25	3.98	0.595	MAH 5– 50 mol%, B	11
P (S-AN)	DMF	20	1.45	0.71	12% S, D	12
			1.80	0.71	25% S	
			2.65	0.72	52% S	
		25	1.86	0.690	15–82% AN, E	13
	THF		1.98	0.681	0–35% AN, E	
	MEK		3.30	0.619	10–48% AN, E	
P (AN-EVE)	DMF	55	0.142	0.9636	18% EVE, F	14
P (SA-AA)	0.5 M NaCl	25	0.81	0.789	10% SA, G, pH 9	15
	1 M NaCl		0.307	0.860		
	0.5 M NaCl		1.09	0.775	20% SA	
	1 M NaCl		1.307	0.759		
	0.5 M NaCl		5.78	0.655	50% SA	
	1 M NaCl		2.98	0.695		
	0.5 M NaCl		0.638	0.844	70% SA	
	1 M NaCl		0.026	1.072		

Abbreviations. THF – tetrahydrofuran; ODCB – o-dichlorobenzene; DEB – diethylbenzene; MEK – methyl ethyl ketone; ns – not specified; PVC – polyvinyl chloride; PVAc – polyvinyl acetate; EPM – ethylene–propylene copolymer; EPDM – terpolymer of ethylene, propylene, diene monomers; EPR – ethylene-propylene rubber; E-NORB – ethylene-norboarnene (50%); EVA – ethylene–vinyl acetate copolymer; S – polystyrene (PS); MMA – polymethyl methacrylate (PMMA); MAH – maleic anhydride; AN – polyacrylonitrile (PAN); EVE – ethyl vinyl ether; SA – sodium acylate; AA – acrylamide.

Remarks. A – use $K = 1.112 \times 10^{-2}$ and $a = 0.723$ for PS; B – $[\eta] = KM_w^a$; C – use $K = 1.55 \times 10^{-2}$ and $a = 0.692$ for PS; D – use $K = 2.40 \times 10^{-2}$ and $a = 0.63$ for PS and $K = 3.25 \times 10^{-2}$ and $a = 0.725$ for PAN; E – $[\eta] = K(W_{AN} + 1.041)^a M_w^a$, W_{AN}: weight fraction of AN unit; F – 0.05 M LiBr included, use $K = 2.2 \times 10^{-2}$ and $a = 0.615$ for PS; G – $[\eta] = KM_v^a$; H – use $K = 1.17 \times 10^{-2}$ and $a = 0.717$ for PS.

References

1. JANČA J, MRKVIČKOVA L, KOLÍNSKY M (1978) J Appl Polym Sci 22:2661
2. SMITH WV (1974) J Appl Polym Sci 18:3685
3. CHIANTORE O, CINQUINA P, GUAITA M (1994) Eur Polym J 30:1043
4. GIANOTTI G, CICUTA A, ROMANINI D (1980) Polymer 21:1087
5. ATKINSON CML, DIETZ R (1976) Makromol Chem 177:213
6. DE CHIRICO A, ARRIGHETTI S, BRUZZONE M (1981) Polymer 22:529
7. BRAUER E, WIEGLEB H, HELMSTEDT M (1986) Polym Bull 15:551
8. ECHARRI J, IRUIN JJ, GUZMAN GM, AMSORENA J (1979) Makromol Chem 180:2749
9. KOTAKA T, TANAKA T, ONHUMI H, MURAKAMI Y, INAGAKI H (1970) Polym J 1:245
10. GOLDWASSER JM, RUDIN A (1983) J Liq Chromatogr 6:2433
11. CHOW CD (1976) J Appl Polym Sci 20:1619
12. KRANZ D, POHL HU, BAUMANN H (1972) Angew Makromol Chem 26:67
13. MENDELSON RA (1987) In: PROVDER T (ed) ACS Symposium Series No 352, ACS, Washington, p 263
14. KENYON AS, MOTTUS EH (1974) Appl Polym Symp 25:57
15. MCCARTHY KJ, BURKHARDT CW, PARAZAK DP (1987) J Appl Polym Sci 33:1699

Appendix IV

Specific Refractive Index Increments

Sample	Solvent	Wave-length (nm)	dn/dc (ml/g)	Temp (°C)	A_2 ($\times 10^3$)	Ref, Remark
Polystyrene	toluene	632.8	0.110	23		1
		632.8	0.108	25		2, A
		632.8	0.108			3
		632.8	0.107			4
		632.8	0.109	22	0.5–0.3	5
		632.8	0.111	20		6
		670	0.108			7
		589.3	0.107			8, C
		546.0	0.104			9
	benzene	546.0	0.1064			9
		546.0	0.109	25		10
		436	0.112	25		10
	CCl_4	546.0	0.146			9
	tetrahydrofuran (THF)	488	0.199			10
		632.8	0.190			3
		632.8	0.193	20		6
		632.8	0.1845	25		11
		632.8	0.184	30		12
		632.8	0.186			4
		632.8	0.184			13
		632.8	0.188	22	0.47–0.3	5
		632.8	0.192	25		14
		632.8	0.184	40		15
		670	0.180	25		11
		670	0.1805	30		16, D
	chloroform	632.8	0.155			3
		632.8	0.1608			17
		546.0	0.169	25		10
		436	0.161	25		10
	acetone	670	0.224			7
	methyl ethyl ketone (MEK)	632.8	0.212	22	0.125–0.06	5
		632.8	0.214			7
		670	0.213			7
		589.3	0.212			8, C
		546.0	0.220			9
	tetrahydronaphthalene	670	0.0615			7

Sample	Solvent	Wave-length (nm)	dn/dc (ml/g)	Temp (°C)	A_2 (× 10³)	Ref, Remark
Polystyrene	MC/DCAA (8:2)	632.8	0.156			18, E
	ethyl benzoate	632.8	0.103	15		19
	dimethyl-	632.8	0.165	20		7
	formamide (DMF)	632.8	0.159			20, F
	tetrachloroethylene (TCE)	632.8	0.0934	25		21
	p-dioxane	546.0	0.152	35		22
(star)		546.0	0.168	35		22
	cyclohexane	546.0	0.170	35		22
(star)		546.0	0.168	35		22
Poly(dimethylsiloxane)	TCE	632.8	−0.0932	25		21
	toluene	632.8	−0.0913	26		20
		632.8	−0.0648	26		20
Poly(styrene-divinylbenzene)	DMF/LiBr	546	0.170	30		23
	THF	546.1	0.1963	30		24
Poly(styrene-methyl methacrylate)	ethyl benzoate	632.8	0.0398	15		19, G
Poly(methyl methacrylate)	THF	632.8	0.083	25	0.2	25
		632.8	0.084			26, H
		632.8	0.084			13
		632.8	0.084			14
		632.8	0.084			27
		632.8	0.09			28
		589	0.0865	23		29
		546	0.0871	25		10
		436	0.0887	25		10
	HFIP	632.8	0.190	25		30, J
		632.8	0.190			27
	toluene	546	0.0157	25		10
		488	0.010	18		19
		439	0.0094	25		10
	DMF	632.8	0.053	25		31
	MEK	546	0.1140			9
		546	0.1173	25		10
		436	0.100	25		10
	benzene	546.0	−0.010			9
		546.0	−0.011	25		10
		436	−0.0039	25		10
	CCl$_4$	546.0	0.023			9
	acetone	546.0	0.1340			9
Poly(ethyl methacrylate)	MEK	546.0	0.104	23		10
	ethyl acetate	436	0.109	25		10
Poly(n-butyl methacrylate)	MEK	546.0	0.102	25		10
		436	0.103	25		10
	benzene	546.0	−0.014	25		10
		436	−0.023	25		10

Sample	Solvent	Wave-length (nm)	dn/dc (ml/g)	Temp (°C)	A_2 ($\times 10^3$)	Ref, Remark
Poly(methacrylic acid)	ethanol	436	0.154	25		10
	methanol	546	0.134	25		10
Poly(methyl acrylate)	benzene	436	−0.019	30		10
	acetonitrile	546.0	0.118	25		10
		436	0.120	25		10
Poly(ethyl acrylate)	acetone	546.0	0.107	23		10
		436	0.109	23		10
	chloroform	546.0	0.0363	25		10
	MEK	546.0	0.0856	20		10
		436	0.0870	20		10
Poly(n-butyl acrylate)	acetone	436	0.0870	20		10
	benzene	546	−0.0292	30		10
	THF	546	0.0651	30		10
		632.8	0.067			6
	toluene	546	−0.0239	30		10
		632.8	−0.024	20		6
	MEK	632.8	0.097	20		6
	DMF	632.8	0.016	20		6
Poly($tert$-butyl acrylate)	MEK	546	0.0818	25		10
	propyl alcohol	546	0.074			32
Poly(vinyl chloride)	THF	632.8	0.101			20
		589.0	0.106	23		29
		589.0	0.115			29
		589.0	0.115	25		10
		578	0.1010	16.5		10
		546.0	0.1017	16.5		10
		546.0	0.116	25		10
		435.8	0.115	20		33
		435.8	0.1124	20		34
		435.8	0.1074	25		35
		435.8	0.104	16.5		10
		435.8	0.119	25		10
	cyclohexanone	632.8	0.0650			20
		589.0	0.076	25		10
		546.0	0.077	25		10
		435.8	0.079	25		10
Polyisoprene	toluene	632.8	0.031			3
	THF	632.8	0.127			3
		589.0	0.156			29
	chloroform	632.8	0.093			3
(cis-1,4)	benzene	546.0	0.0194	20		10
		436	0.0143	20		10
($trans$-1,4)	cyclohexane	436	0.117			10
Polybutadiene	toluene	632.8	0.032			3
	THF	632.8	0.132			3
		632.8	0.1295	26		20, K
		589.0	0.132	40		29
	chloroform	632.8	0.094			3
		546.0	0.083			10

Sample	Solvent	Wave-length (nm)	dn/dc (ml/g)	Temp (°C)	A_2 (× 10³)	Ref, Remark
Polybutadiene						
(*cis*-1,4)	cyclohexane	632.8	0.1151	30		36
(*trans*-1,4)		632.8	0.1084	30		36
(1,2-)		632.8	0.0873	30		36
		632.8	0.111	30		36, L
		546	0.118			10
		546	0.1174	25		9
		436	0.126			10
		436	0.1345	25		9
	benzene	546	0.0086	20		10
		546	0.0099	25		9
		436	0.0117	20		10
		436	0.0160	25		9
	hexane	546	0.1523	25		9
		436	0.2181	25		9
Polychloroprene	THF	632.8	0.1274	25		11
	MEK	546	0.156	25		10
		436	0.162	25		10
Polyisobutylene	1-chloronaphthalene	632.8	−0.122	135		37
		578	−0.128	135		37
		546	−0.131	135		37
		436	−0.146	135		37
	trichlorobenzene	632.8	−0.051	135		37
	(TCB)	578	−0.053	135		37
		546	−0.054	135		37
		436	−0.066	135		37
	THF	632.8	0.1026	23		20, M
		632.8	0.125	25	0.35	38
		546.1	0.114	25		39
	toluene	632.8	0.106	25		2, B
Poly(1-butene)	1-chloronaphthalene	632.8	−0.177	135		37
		578	−0.183	135		37
		546	−0.187	135		37
		436	−0.206	135		37
	TCB	632.8	−0.103	135		37
		578	−0.105	135		37
		546	−0.108	135		37
		436	−0.120	135		37
	n-nonane	436	−0.108	35		10
Polyethylene	TCB	632.8	−0.097	135		20
		632.8	−0.104	135		12
		632.8	−0.098	145		12
		632.8	−0.098	145		40
		632.8	−0.091	145		41
		546	−0.106	135		10
(MW 11 000)		589.3	−0.1063	135		8, C
(MW 18 400)		589.3	−0.1085			8, C
(MW 11 000)	1-chloronaphthalene	632.8	−0.177	145		20

Sample	Solvent	Wave-length (nm)	dn/dc (ml/g)	Temp (°C)	A_2 ($\times 10^3$)	Ref, Remark
Polyethylene		589.3	−0.1883			8
		546.0	−0.191	135		42
	decane	577	0.1200	125		43
		546	0.1213	125		43
		436	0.1263	125		43
		404	0.1233	125		43
	o-dichlorobenzene	632.8	−0.056	135		20
	(ODCB)	589	−0.090	135		29
		546	−0.078	120		10
	1-cyclonaphthalene	546	−0.214	125		10
		436	−0.245	125		10
HDPE	TCB	632.8	−0.112	145	2.00	44
		632.8	−0.107	135		37
		578	−0.108	135		37
		546	−0.110	135		37
		436	−0.125	135		37
(SRM 1475)		632.8	−0.104	135	1.90	45
	ODCB	632.8	−0.056	145	1.41	44
		589	−0.069	135		29
(SRM 1475)		632.8	−0.056	135		45
	1-chloronaphthalene	632.8	−0.189	145	0.68	44
		632.8	−0.183	135		37
		578	−0.188	135		37
		546	−0.192	135		37
		436	−0.215	135		37
(SRM 1475)		632.8	−0.177	145		45
(SRM 1475)		589.3	−0.1932	135		8
	diphenylmethane	546	−0.125	142		46
LDPE	TCB	632.8	−0.108	145	3.05	44
		632.8	−0.102	135		37
		578	−0.103	135		37
		546	−0.105	135		37
		436	−0.117	135		37
(SRM 1476)		632.8	−0.091	135	4.3	45
		632.8	−0.091	140	0.42	47
	ODCB	632.8	−0.051	145	1.71	44
		589.0	−0.104	138		29
	1-chloronaphthalene	632.8	−0.185	145	0.87	44
		632.8	−0.178	135		37
		578	−0.184	135		37
		546	−0.189	135		37
		546	−0.195	125		48
		436	−0.212	135		37
(SRM 1476)		589.3	−0.1916	135		8
Linear-LDPE	TCB	632.8	−0.106	145	1.50	44
		632.8	−0.106	145		49, N
		632.8	−0.103	145		49, O
	ODCB	632.8	−0.048	145	1.05	44
	1-chloronaphthalene	632.8	−0.180	145	0.79	44

Sample	Solvent	Wave-length (nm)	dn/dc (ml/g)	Temp (°C)	A_2 (× 10³)	Ref, Remark
Polypropylene	1-chloronaphthalene	546	−0.227	125		10
		546	−0.191	125		27
		436	−0.228	125		10
(isotactic)		632.8	−0.173	135		37
(isotactic)		578	−0.180	135		37
(isotactic)		546	−0.184	135		37
(isotactic)		436	−0.205	135		37
(atactic)		632.8	−0.177	135		37
(atactic)		578	−0.183	135		37
(atactic)		546	−0.188	135		37
(atactic)		436	−0.211	135		37
	TCB	632.8	−0.095	135		50
		632.8	−0.093	145		49
		632.8	−0.091	145		20
(MW 235 800)		632.8	−0.092	145		51
(MW 175 100)		632.8	−0.094	145		51
(isotactic)		632.8	−0.102	135		37
(isotactic)		578	−0.105	135		37
(isotactic)		546	−0.107	135		37
(isotactic)		436	−0.121	135		37
(atactic)		632.8	−0.105	135		37
(atactic)		578	−0.109	135		37
(atactic)		546	−0.111	135		37
(atactic)		436	−0.123	135		37
Polyoxymethylene	DMF	632.8	−0.0505	135		52
Polyethylene terephthalate	HFIP	632.8	0.257		2.9	53
	HFIP/PFP (5:5)	632.8	0.242			53, P
	MC/DCAA (8:2)	632.8	0.148	RT		18, E
	DCAA	589	0.106			54, Q
	m-cresol	589.0	−0.073	25		29
	o-chlorophenol	589.0	−0.070	25		29
Polyamide 6	HFIP	632.8	0.2375	25		55
	MC/DCAA (8:2)	632.8	0.180	RT		56
	o-chlorophenol	589.0	−0.012	90		29
Polyamide 11	MC/DCAA (8:2)	632.8	0.133	RT		56
	HFIP	546	0.25	30	7.8 − 5.6	57
Polyamide 12	MC/DCAA (8:2)	632.8	0.133	RT		56
	HFIP	632.8	0.2212			58
		546	0.25	30		59
Polyamide 6, 6	MC/DCAA (8:2)	632.8	0.182	RT		56
(46 K)			0.188	RT		56
(32 K)			0.194	RT		56
	HFIP	632.8	0.241	25	4.8	60
	formic acid	632.8	0.136	25		60
	m-cresol	589.0	−0.016	130		29
Polyamide 6, 9	MC/DCAA (8:2)	632.8	0.163	RT		56

Sample	Solvent	Wave-length (nm)	dn/dc (ml/g)	Temp (°C)	A_2 ($\times 10^3$)	Ref, Remark
Polyamide 6, 10	MC/DCAA (8:2)	632.8	0.163	RT		56
Polyamide 6, 12	MC/DCAA (8:2)	632.8	0.149	RT		56
Polyamide 6T	MC/DCAA (8:2)	632.8	0.201	RT		56
Silk (Silkworm)	HFIP	632.8	0.236			61
(Spider)	HFIP	632.8	0.235			61
Polyacrylonitrile	DMF	632.8	0.092	25		31
		546	0.082	20		10
		436	0.087	20		10
	dimethylsulfoxide	546	0.031	25		10
		436	0.029	25		10
	dimethylacetamide	546	0.0769	25		10
		436	0.0767	25		10
Poly(acrylonitrile-styrene) (28.5/71.5)	THF	632.8	0.1546	23		20
Polyurethane	THF	632.8	0.0863	25		62
		632.8	0.1469	23		20, R
Poly(methyl vinyl ether)	cyclohexane	632.8	0.050	50		63
PMVEMA	0.1 M tris	632.8	0.227			64, S
Poly(phenyl acetylene)	THF	632.8	0.2864	25		65
Epoxy resin	THF	632.8	0.187	23		20
Phenoxy resin	THF	632.8	0.1773			26
Polycarbonate	THF	632.8	0.177			20
		632.8	0.1855			20
Polyphosphazene I	THF	632.8	0.160	30		66, T
II		632.8	0.145	30		66, T
III		632.8	−0.029	30		66, T
IV		632.8	−0.023	25	1.0	67, T
	acetone	632.8	0.019	25	5.1	67, T
	cyclohexanone	632.8	−0.053	40	6.7	67, T
Poly(ether sulfone)	DMF	632.8	0.188	25		4
Polyimide	DMF	632.8	0.208	25		4
	N-methylpyrrolidone	632.8	0.132			68, U
PEEK (MW 14300–22600)	conc. H_2SO_4	632.8	0.376		4.3–2.0	69, V
(MW 28300–56500)		632.8	0.419		1.8–2.7	69, V
Poly(phenylene sulfide)						
	1-chloronaphthalene	632.8	0.167	220		70
Poly(vinyl butyral)	HFIP	632.8	0.189	25		30, J
	THF	632.8	0.089			30, J

Sample	Solvent	Wave-length (nm)	dn/dc (ml/g)	Temp (°C)	A_2 (× 10³)	Ref, Remark
Poly(2-vinylpyridine)	benzene	546	0.075	25		71
Poly(4-vinylpyridine)	DMF	436	0.160	25		10
	chloroform	436	0.150	25		10
Poly(vinylpyrrolidone)	DMF	632.8	0.095	23		20
	chloroform	546	0.108			10
		436	0.108			10
	water	546	0.185			10
		436	0.185			10
6 FDA/p-PDA	N-methylpyrrolidone	632.8	0.107			72, W
6 FDA/MDA		632.8	0.132			72, W
BTDA/DDS		632.8	0.145			72, W
Methylated humic acid	THF	632.8	0.22		−0.042	73
Benzylated humic acid	THF	632.8	0.19		−0.041	73
Hydroxyethyl cellulose acetate	THF	488	0.0734	25		74
Poly(vinyl acetate)	THF	632.8	0.0517	25		11
		632.8	0.0471	26		20, X
		632.8	0.0402	26		20, Y
	ethyl acetate	632.8	0.077	23		20
	acetone	546	0.104	25		10
		436	0.104	25		10
	MEK	546	0.080	25		10
		436	0.083	25		10
Poly(vinyl alcohol)	0.05 M NaNO₃	632.8	0.1501	27		75
	0.05 M NaNO₃	632.8	0.1429	27		75, Z
		632.8	0.1501	25		76, a
		632.8	0.1429	25		76, b
	0.1 M KH₂PO₄	632.8	0.1534	23		20
	water	546	0.164	30		10
		436	0.168	30		10
Polyacrylamide	water	632.8	0.1735			77
		632.8	0.142	23		20, C
		632.8	0.1829	25		20, d
		546	0.163			10
	1 M NaCl	632.8	0.1587	25		20, e
	1 M NaCl	632.8	0.1556	25		20, f
	0.1 M Na₂HPO₄	632.8	0.163			78
	0.02 M Na₂SO₄	632.8	0.176	23		79, g
			0.174	23		79, h
Poly(acrylamide-co--sodium acrylate)	1 M NaCl	632.8	0.165			80
	0.1 M NaH₂PO₄ + 0.3 M Na₂SO₄, pH 2.2					
		632.8	0.16			81
(70/30)	0.1 M NaNO₃	632.8	0.165			20
Sodium polyacrylate	0.3 M NaCl	488	0.159	25		82
	0.1 M NaCl	632.8	0.175	23		20
	water	632.8	0.1463	23		20, i

Sample	Solvent	Wave-length (nm)	dn/dc (ml/g)	Temp (°C)	A_2 ($\times 10^3$)	Ref, Remark
Poly(acrylic acid)	aqueous NaCl	436	0.158	30		10
	THF	632.8	0.0738	23		20
		632.8	0.0994	23		20
	0.2 N HCl	632.8	0.133			83
Poly(allyl amine)	0.1% TFA/	632.8	0.2034	35		84, j
	0.20 M NaNO$_3$	632.8	0.1950	35		84, k
Poly(vinyl amine)	0.1% TFA/0.20 M NaNO$_3$	632.8	0.1751	35		84, m
Branched polyethyleneimine						
	water	632.8	0.210	35		85
	1 M NaCl	632.8	0.195	35		85
	methanol	632.8	0.225	35		85
Poly(ethylene oxide)	methanol	632.8	0.145	25		86
Dextran	water	632.8	0.147			87, n
	water	632.8	0.150	25		88
	0.1 M KNO$_3$	632.8	0.150	25		88
	0.05 M KH$_2$PO$_4$ (pH 7)	632.8	0.137			87, n
	0.15 M NaCl	632.8	0.145			89
	0.05 M K$_2$HPO$_4$ (pH 7)	632.8	0.1378		0.41	90
	water	546	0.148			91
	0.5 M NaCl	546	0.145			91
	water	436	0.150			91
	0.5 M NaCl	436	0.148			91
	0.2 M NaNO$_3$	514.5	0.147	RT		92
DEAE-dextran	0.8 M NaNO$_3$	632.8	0.150	20		93, o
Pullulan	0.2 M NaNO$_3$	514.5	0.147	RT		92
	0.15 M NaCl	632.8	0.145			89
	water	488	0.148	25		94
Polysaccharide	water	632.8	0.146			95
	0.1 M NaCl	632.8	0.143			96, p
(anionic)	0.1 M NaCl	632.8	0.143			96, q
Sodium hyaluronate	0.1 M NH$_4$NO$_3$	632.8	0.155		3	97
Sodium heparin	0.1 M NaCl	632.8	0.163	23		98, r
Sodium poly(styrene sulfonate)						
	water	632.8	0.17			99
(0.9 meq/g)	0.05 M NaCl	546	0.212	25		10
(1.8 meq/g)		546	0.207	25		10
(2.0 meq/g)	0.10 M NaCl	546	0.195	25		10
(2.7 meq/g)		546	0.205	25		10
	0.1 M NaNO	632.8	0.195			100
Poly(sulfoalkyl methacrylate)						
(ethyl)	0.1 M NaCl	532	0.135	25		101
(hexyl)		532	0.140	25		101
(decyl)		532	0.170	25		101
(octyl)		532	0.149	25		101
Hydroxyethyl-cellulose	0.1 M NaCl	632.8	0.159			102

Sample	Solvent	Wave-length (nm)	dn/dc (ml/g)	Temp (°C)	A_2 ($\times 10^3$)	Ref, Remark
Carboxylmethyl-cellulose	0.1 M NaCl	632.8	0.147			102
PNaAMPS	water	632.8	0.20			99, s
BSA	water	632.8	0.187	25		103, t
BSA-SDS	water	632.8	0.376	25		103, u
Ovalbumin-SDS	water	632.8	0.374	25		103
Ribonuclease-SDS	water	632.8	0.398	25		103
Ribonuclease A	75%A – 25%B	632.8	0.172		7.5×10^{-2}	104, v
Lysozyme	65%A – 35%B	632.8	0.182		-5.39×10^{-2}	104, v
BSA	58%A – 42%B	632.8	0.172		7.49×10^{-2}	104, v
Bovin alkaline phosphatase	50%A – 50%B	632.8	0.142		0.854	105, w
	60%A – 40%B	632.8	0.1188		–9.81	105, w
	70%A – 30%B		0.0956		–163	106, w
	70%A – 30%B	632.8	0.0956		–163	105, w
Bovin thyroglobulin (MW 669000)	0.10 M Na$_2$SO$_4$	632.8	0.165	20		106, x
Serum albumin (MW 68000)	0.10 M Na$_2$SO$_4$	632.8	0.165	20		106, x
Chymotrypsinogen A (MW 25900)	0.10 Na$_2$SO$_4$	632.8	0.170	20		106, x
DNA	water	632.8	0.166			99
DNA fragment	0.1 M NaCl	632.8	0.168	25	0.6	107

Remarks. A: dn/dc = 0.108 – 21.0/M_n; B: dn/dc = 0.106 – 12.4/M_n; C: polystyrene MW-(dn/dc)$_{toluene}$-(dn/dc)$_{MEK}$, 1.8×10^6–0.109–0.216, 97200–0.107–0.214, 51000–0.106–0.212, 19800–0.103–0.213, 10300–0.102–0.209, 4800–0.101–0.210, 2030–0.0985–0.208, polyethylene/trichlorobenzene dn/dc = –(0.1023 + (20.4/M)); D: (dn/dc)$_{670}$ = 0.9783 (dn/dc)$_{632.8}$, (dn/dc)$_{670}$ = 0.1804 – 9.149/M; E: methylene chloride; F: DMF – N,N-dimethylformamide; G: 60% MMA diblock; H: PS/THF, A_2 = 0.01 $M_w^{-0.25}$, PMMA/THF, A_2 = 0.012 $M_w^{-0.29}$; J: vinyl alcohol content ranging from 11 to 23%, hexafluoroisopropanol; K: M_w = 6160; L: dn/dc = 0.111 – 15.3/M_n (36% cis-1,4, 57% $trans$-1,4, 7% 1,2-); M: M_w = 6100 (0.0974 for M_w = 3300); N: ethylene-co-octene-1; O: ethylene-co-butene-1; P: pentafluorophenol; Q: dichloroacetic acid; R: M_w – dn/dc, 35500 – 0.1417, 46200 – 0.1523, 63100 – 0.1477, 72400 – 0.1469; S: poly(methyl vinyl ether-co-maleic anhydride), hydroxymethyl aminomethane, pH 9; T: I – poly (diphenoxy)phosphazene, II – poly(aryloxy)phosphazene, III – poly(fluoroalkoxy)phosphazene, IV – poly[bis(trifluoroethyl) phosphazene]; U: 6F-dianhydride-1,4-phenylenediamine dn/dc = 0.107, 6F-dianhydride-4,4-methylenediamine dn/dc = 0.132, benzophenone-3,3′,4,4′-diaminodiphenylsulfone dn/dc = 0.145; V: poly(arylether ether ketone); W: 6FDA – 6F-dianhydride, MDA – 4,4′-methylenediamine, DDS – 4,4′-diamine diphenyl sulfone, BTDA – benzophenone-3,3′,4,4′-tetracarboxylic dianhydride, PDA – 1,4-phenylenediamine; X: M_w = 4.66 × 10^6; Y: M_w = 1.07 × 10^5; Z: 88% degree of hydrolysis; a: 98% hydrolysis; b: 88% hydrolysis; c: M_w = 5.5 × 10^6; d: M_w = 3.69 × 10^6; e: M_w = 7.72 × 10^5; f: M_w = 8 × 10^6; g: 5.32 × 10^5 M_w; h: 2.18 × 10^6 M_w; i: M_w = 28000; j: TFA – trifluoroacetic acid, M_w = 3.61 × 10^4; k: M_w = 1.06 × 10^5; m: M_w = 2.50 × 10^5; n: M_w = 6.82 × 10^4; o: DEAE – 2-diethylaminoethyl; p: Schizophyllan (neutral polysaccharide); q: xanthan (anionic heteropolysaccharide); r: 0.01 M NaCl – 0.142 – A_2 = 16, 0.1 M NaCl – 0.137 – 5.8, 0.5 M NaCl – 0.126 – 2.52; 1 M NaCl – 0.121 – 1.52; s: poly(sodium-2-(acrylamide)-2-methyl propanesulfonate); t: bovine serum albumin; u: sodium dodecyl sulfonate; v: A – 0.15 % TFA in water, B – 0.15 % TFA in 95 % acetonitrile; w: A – 2.5 M ammonium sulfate in 0.5M ammonium acetate (pH 6.0), B – 0.5M ammonium acetate (pH 6.0); x: in 0.02 M Na_2HPO_4 – NaH_2PO_4 at pH 6.9.

References

1. ANDERSON RI (1976) Am Lab 8(5):95
2. HADJICHRISTIDIS N, FETTERS LJ (1982) J Polym Sci Polym Phys Ed 20:2163
3. JORDAN RC, SILVER SF, SEHON RD, RIVARD RJ (1984) In: PROVDER T (ed) Size Exclusion Chromatography, ACS Symposium Series 245, American Chemical Society, Washington, DC, p 295
4. SEGUDOVIC N, KARASZ FE, MACKNIGHT WJ (1990) J Liq Chromatogr 13:2581
5. YUNAN W, ZHONGDE X, LI J, ROSENBLUM WM, MAYS JW (1993) J Appl Polym Sci 49:967
6. MALIHI FB, KUO CY, PROVDER T (1984) J Appl Polym Sci 29:925
7. FRANK R, FRANK L, FORD NC (1995) In: PROVDER T, BARTH HG, URBAN MW (eds) Chromatographic Characterization of Polymers, Adv Chem Ser 247, American Chemical Society, Washington, DC, p 110
8. WAGNER HL, HOEVE CAJ (1971) J Polym Sci A-2, 9:1763
9. KHAN HU, GUPTA VK, YAMIN M (1983) J Macromol Sci – Chem A20:503
10. BRANDRUP J, IMMERGUT EH (1972) Polymer Handbook, Wiley, New York
11. JORDAN RC, MCCONNELL ML (1980) In: PROVDER T (ed) Size Exclusion Chromatography, ACS Symposium Series 138, Americal Chemical Society, Washington, DC, p 107
12. JENG L, BALKE ST, MOUREY TH, WHEELER L, ROMEO P (1993) J Appl Polym Sci 49:1359
13. MOUREY TH, MILLER SM, BALKE ST (1990) J Liq Chromatogr 13:435
14. LEE HC, REE M, CHANG T (1995) Polymer 36:2215
15. JENG L, BALKE ST (1993) J Appl Polym Sci 49:1375
16. MOUREY TH, COLL H (1995) In: PROVDER T, BARTH HG, URBAN MW (eds) Chromatographic Characterization of Polymers, Adv Chem Ser 247, American Chemical Society, Washington, DC, p 123
17. OUANO AC, KAYE W (1974) J Polym Sci Polym Chem Ed 12:1151
18. MOUREY TH, BRYAN TG, GREENER J (1993) J Chromatogr A 657:377
19. KENT MS, TIRRELL M, LODGE TP (1994) J Polym Sci Part B Polym Phys Ed 32:1927 (1994)
20. Wyatt Technology, Data Sheet, Santa Barbara, CA
21. DUMELOW T (1989) J Macromol Sci-Chem A26:125
22. SUN SF, CHAO D, LIN S (1989) Polym Commun 24:84
23. FORGET JL, BOOTH C, CANHAM PH, DUGGLEBY H, KING TA, PRICE C (1979) J Polym Sci Polym Phys Ed 17:1403
24. AMBLER MR, MCINTYRE D (1977) J Appl Polym Sci 21:3237
25. LEE S, KWON O-S (1995) In: PROVDER T, BARTH HG, URBAN MW (eds) Chromatographic Characterization of Polymers, Adv Chem Ser 247, Americal Chemical Society, Washington, DC, p 93
26. PODZIMEK S (1994) J Appl Polym Sci 54:91
27. BERKOWITZ SA (1983) J Liq Chromatogr 6:1359 (1983)
28. DEGOULET C, NICOLAI T, DURAND D, BUSNEL JP (1995) Macromolecules 28:6819
29. EVANS JM (1973) Polym Eng Sci 13:401
30. REMSEN EE (1991) J Appl Polym Sci 42:503
31. ASADUZZAMAN AKM, RAKSHIT AK, DEVI S (1993) J Appl Polym Sci 47:1813
32. MRKVICKOVA J, DANHELKA J (1990) J Appl Polym Sci 41:1929
33. HATTORI S, ENDOH H, NAKAHARA H, KAMATA T, HAMASHIMA M (1978) Polym J 10:173
34. RUDIN A, BENSCHOP-HENDRYCHOVA I (1974) J Appl Polym Sci 15:2881 (1971)

35. GEERISSEN H, ROOS J, WOLF BA (1985) Makromol Chem 186:735
36. CHEN X, XU Z, HADJICHRISTIDIS N, FETTERS LJ, CARELLA J, GRAESSLEY WW (1984) J Polym Sci Polym Phys Ed 22:777
37. HORSKA J, STEJSKAL J, KRATOCHVIL P (1983) J Appl Polym Sci 28:3873
38. HUBER CH, LEDERER K (1980) J Polym Sci Polym Let Ed 18:535
39. MRKVICKOVA L, LOPOUR P, POKORNY S, JANCA J (1980) Angew Makromol Chem 90:217
40. GRINSHPUN V, O'DRISCOLL KF, RUDIN A (1984) J Appl Polym Sci 29:1071
41. PANG S, RUDIN A (1993) In: PROVDER T (ed) Chromatography of Polymers, ACS Symposium Series 521, American Chemical Society, Washington, DC, p 254
42. WILD L, RANGANATH R, RYLE T (1971) J Polym Sci A-2 9:2137
43. BOEHN LL, LANVER U, LECHNER MD (1983) Makromol Chem 184:585
44. GRINSHPUN V, O'DRISCOLL KF, RUDIN A (1984) In: PROVDER T (ed) Size Exclusion Chromatography, ACS Symposium Series 245, American Chemical Society, Washington, DC, p 273
45. MACRURY TB, MCCONNELL ML (1979) J Appl Polym Sci 24:651
46. HORSKA J, STEJSKA J, KRATOCHVIL P (1979) J Appl Polym Sci 24:1845
47. RAM A, MILTZ J (1971) J Appl Polym Sci 35:2639
48. DROTT EE, MENDELSON RA (1970) J Polym Sci Part A-2, 8:1373
49. GRINSHPUN V, RUDIN A (1985) Makromol Chem Rapid Commun 6:219
50. BILLIANI J, LEDERER K (1990) J Liq Chromatogr 13:3013
51. GRINSHPUN, RUDIN A (1985) J Appl Polym Sci 30:2413
52. OGAWA T (1990) J Liq Chromatogr 13:51
53. BERKOWITZ S (1984) J Appl Polym Sci 29:4353
54. OHOYA S, MATSUO T (1989) Chem Expr (JPN) 4:761
55. SCHORN H, KOSFELD R, HESS M (1983) J Chromatogr 282:579
56. MOUREY TH, BRYAN TG (1994) J Chromatogr A 679:201
57. TERAN CC, MACCHI EM, FIGINI RV (1987) J Appl Polym Sci 34:2433
58. OGAWA T, SAKAI M (1985) J Polym Sci Polym Chem Ed 23:1109
59. HAMMEL R, GERTH CH (1977) Makromol Chem 178:2697
60. SAMANTA SR (1992) J Appl Polym Sci 45:1635
61. JACKSON C, O'BRIEN JP (1995) Macromolecules 28:5975
62. PETROVIC ZS, MACKNIGHT WJ (1991) Polym Bull 27:281
63. PETRI HM, STAMMER A, WOLF BA (1995) Makromol Chem Phys 196:1453
64. WU CS, SENKA L, MALAWER EG (1989) J Liq Chromatogr 12:2919
65. SEDLACK J, VOHLIDE J (1993) Makromol Chem Rapid Commun 14:51
66. DE JAEGER R, LECACHEUX D, POTIN PH (1990) J Appl Polym Sci 39:1793
67. MOUREY TH, MILLER SM, FERRAR WT, MOLAIRE TR (1989) Macromolecules 22:4286
68. KONAS M, MOY TM, ROGERS ME, SCHULTZ AR, WARD TC, MCGRATH JE (1995) J Polym Sci Part B Polym Phys 33:1441
69. DEVAUX J, DELIMOY D, DAOUST D, LEGRAS R, MERCIER JP, NIELD E (1985) Polymer 26:1994
70. STACY CJ (1986) J Appl Polym Sci 32:3959
71. TALLEY CP, BOWMAN LM (1979) Anal Chem 51:2239
72. KONAS M, MOY TM, ROGERS ME, SHULTZ AR, WARD TC, MCGRATH JE (1995) J Polym Sci Part B Polym Phys 33:1441
73. OLSON ES, DIEHL JW (1985) J Chromatogr 349:337
74. WU C, SIDDIG M, JIANG S, HUANG Y (1995) J Appl Polym Sci 58:1779
75. NAGY DJ (1995) In: WU C-S (ed) Handbook of Size Exclusion Chromatography, Dekker, New York, p 279
76. NAGY DJ (1986) J Polym Sci Part C Polym Let Ed 24:87
77. LESEC J, VOLET G (1989) International GPC Symposium, Massachusetts, p 386
78. LANGHORST MA, STANLEY FW JR, CUTIE SS, SUGARMAN JH, WILSON LR, HOAGLAND DA, PRUD'HOMME RK (1986) Anal Chem 58:2242
79. KIM CJ, HAMIELEC AE, BENEDEK A (1982) J Liq Chromatogr 5:1277
80. MULLER G, YONNET C (1984) Makromol Chem Rapid Commun 5:197

81. PAPAZIAN LA (1990) J Liq Chromatogr 13:3389
82. KATO T, TOKUYA T, NOZAKI T, TAKAHASHI A (1984) Polymer 25:218
83. LACIK I, SELB J, CANDAN F (1995) Polymer 36:3197
84. NAGY DJ, TERWILLIGER DA, LAWREY BD, TIEDGE WF (1989) International GPC Symposium, Massachusetts, p 637
85. PARK IH, CHOI EJ (1996) Polymer 37:313
86. KINUGASA S, NAKAHARA H, KAWAHARA J, KOGA Y, TAKAYA H (1996) J Polym Sci Part B Polym Phys 34:583
87. LDC/Milton Roy/Chromatix Technical Note LS 7 (1979)
88. VAN DIJK JAPP, VARKEVISSER FA, SMIT JAM (1987) J Polym Sci Part B Polym Phys 25:149
89. JACKSON C, NILSSON L, WYATT PJ (1989) J Appl Polym Sci Appl Polym Symp 43:99
90. KIM CJ, HAMIELEC AE, BENEDEK A (1982) J Liq Chromatogr 5:425
91. VINK H, DAHLSTROM G (1967) Makromol Chem 109:249
92. BAHARY WS, HOGAN MP, JILANI M, ARONSON MP (1995) In: PROVDER T, BARTH HG, URBAN MW (eds) Chromatographic Characterization of Polymers, Adv Chem Ser 247, American Chemical Society, Washington, DC, p 151
93. HANSELMANN R, BURCHARD W (1995) Makromol Chem Phys 196:2259
94. KATO T, OKAMOTO T, TOKUYA T, TAKAHASHI A (1982) Biopolymers 29:1623
95. YU LP, ROLLINGS JE (1988) J Appl Polym Sci 35:1085
96. CAPRON I, GRISEL M, MULLER G (1995) Int J Polym Anal & Charact 2:9
97. REED WF (1995) Makromol Chem Phys 196:1539
98. MIKLAUTS H, RIEMANN J, VIDIC HJ (1986) J Liq Chromatogr 9:2073
99. HOLZWARTH G, SONI L, SCHULZ DN (1986) Macromolecules 19:422
100. TIELKING H, KULICKE WM (1996) Anal Chem 68:1169
101. PEIGIANG W, SIDDIG M, HUIYING C, DI Q, WU C (1996) Macromolecules 29:277
102. PICTON L, MERLE L, MULLER G (1996) Int J Polym Anal & Character 2:103 (1996)
103. KONISHI K (1988) In: DUBIN P (ed) Aqueous Size Exclusion Chromatography, Elsevier, Amsterdam, p 327
104. MHATRE R, KRULL IS, STUTING HH (1990) J Chromatogr 502:21
105. KRULL IR, STUTING HH, KRZYSKO SC (1988) J Chromatogr 442:29
106. BINDELS JG, DE MAN BM, HOENDERS HJ (1982) J Chromatogr 252:255
107. NICOLAI T, VAN DIJK L, VAN DIJK JAPP, SMIT JAM (1987) J Chromatogr 389:286

Subject Index